"Here's a book that will tell you everything you ne
telephone calls over your PC. No one knows voice or.

Josh Quittner, *Time Magazine*

"Jeff Pulver is the Thomas Jefferson of internet telephony: philosopher, architect, advisor, diplomat, inventor, analyst, and guide to all those intrigued by the promise and power of this remarkable field."

Fred Hapgood, Contributor, *Wired*

"Jeff Pulver is the preeminent authority on Internet telephony. He offers deep insights into all aspects of this industry, from the technology to the business part- nerships to the nature of the marketplace. He knows the things that will determine which products will succeed and which will be left by the wayside."

Gary Welz, *Internet World*

"Jeff Pulver is one of the great pioneers of Internet telephony. When the history of communications in the twentieth century is written, he'll merit pages to himself. His explanations of the technology, and the anecdotes he wraps them up in, are always pithy and amusing—and he has a great populariser's touch."

Tim Jackson, *Financial Times*

"Jeff Pulver has consistently been on the leading edge of Internet politics, technol- ogy, and communications. Once upon a time, there was *Steal This Book*. Today, we have Jeff Pulver's version of *Steal This Internet Phone!*"

Robert Knight, Host of the *Free Speech Radio Show* on WBAI, New York City

"If successful, Internet telephony could undermine the economics of the interna- tional telecommunications model. Jeff Pulver has been at the forefront of this move- ment."

Kenneth Hart, *Communications Week*

"Long distance phone companies circle the wagons when Jeff Pulver is on the loose. From Taipei to Tel Aviv, Pulver's known as the most knowledgeable practioner of Internet telephony."

Art Kramer, *The Atlanta Constitution*

"Jeff Pulver's intuitive understanding and passion for Internet telephony is a gift to all of us. Months before even its opponents understood the promise of telephony, Pulver knew its potential. And Pulver can explain it even to me!"

Frank Barnako, CBS Radio Network, Washington, D.C.

"Jeff Pulver is one of those rare citizens that help shape society. In a super-dramatic environment like the Internet, Jeff is the Super-Citizen, devoting endless time and effort to community work. At the beginning of 1995 when VocalTec pioneered Internet Phone and made the free global voice communication idea into reality, Jeff stepped forward as a private citizen to champion this exciting technology."

Elon Ganor, CEO, VocalTec, Inc.

"The Internet is about communication, and it isn't just text anymore! Jeff Pulver has been an innovator and an advocate for the technology that is allowing people all around the world to communicate by voice over the Net. To call him tireless would be an understatement. His knowledge, drive, and enthusiasm are helping to shape the convergence of computers and communication."

Fred Fishkin, Host and Producer of "Bootcamp, a Report on Computers & Technology," heard on CBS Radio.

The Internet Telephone Toolkit

The Internet Telephone Toolkit

Jeff Pulver

WILEY COMPUTER PUBLISHING

John Wiley & Sons, Inc.

New York • Chichester • Brisbane • Toronto • Singapore • Weinheim

Publisher: Katherine Schowalter
Editor: Theresa Hudson
Senior Managing Editor: Frank Grazioli
Electronic Products, Associate Editor: Mike Green
Text Design & Composition: Benchmark Productions, Inc.

Designations used by companies to distinguish their products are often claimed as trademarks. In all instances where John Wiley & Sons, Inc. is aware of a claim, the product names appear in initial capital or all capital letters. Readers, however, should contact the appropriate companies for more complete information regarding trademarks and registration.

Adobe, Acrobat, and Adobe Type Manager are trademarks of Adobe Systems, Inc., and may be registered in certain jurisdictions.

This text is printed on acid-free paper.

Library of Congress Cataloging-in-Publication Data:

ISBN 0 471-16352-X

Printed in the United States of America
10 9 8 7 6 5 4 3 2 1

For Dylan and Jake,
Love,
Dad

Contents

Acknowledgments

This book would not have happened were it not for my editor, Terri Hudson, and her follow-up and persistence. Terri really was an absolute pleasure to work with and I've learned from Terri what it really takes to get a book ready for production.

I'd like to thank the entire team at John Wiley & Sons for working with me to get this book published. In addition to Terri, I'd like to thank senior managing editor, Frank Grazioli, and associate editor Mike Green, who helped master the enclosed CD-ROM.

When I look back at all of the people who influenced me in writing this book, I feel compelled to acknowledge the following people:

My wife, Risa, for her encouragement and being there for me whenever I needed. Risa, thanks for bearing with me and my obsession with the Net.

My grandparents Saul, Tessie, and Shirley for their encouragement.

My parents Jeanne and Howard, for introducing me to computers and ham radio at an early age.

My sisters Lauren and Michele, for bearing with my ham radio indulgences, including those during our family vacations. (CQ DX, CQ DX, CQ DX).

Josh Quittner, who may not realize it, but was the one who first introduced me to CU-SeeMe through his weekly column in *Newsday*. It was here that I learned about the launch of Internet Phone.

Hugh Hamilton, for helping me test some of the telephony products in the book at two in the morning.

Marshall Caro, James Talvy, Ben Lo, and Alex Schmidt, for teaching me everything I needed to know about UDP but was afraid to ask.

Sandy Combs and Bruce Jacobs, for their incredible hard work and effort in the startup of the VON Coalition and their feedback regarding the book.

All of the subscribers to the mailing lists at pulver.com for their continued contributions and support.

All of the readers of the "Pulver Reports" Web pages at pulver.com (http://www.pulver.com)

All the people I've met on ham radio and the Net these past few years who have influenced me in conversations and in e-mail exchanges.

And my sons Dylan and Jake, who were my inspiration for writing this book.

Preface

If you're interested in learning how you can make long distance phone calls over the Internet for *free*, read on. This book gives you the background information to what "Internet phoning" is all about. It also provides detailed information and insider tips on the most popular internet telephony products currently available. I've included a CD-ROM containing many of the popular Internet Telephone software products so you can start calling . . . today!

The Internet has really started to affect our everyday lives. The next time you open a newspaper and look for a current movie, take a look at all of the World Wide Web pages that are referenced in the movie ads. Ready or not, URLs, Web page references, and e-mail addresses are as commonplace as the telephone number.

With the enormous growth of the Internet, new and exciting technologies have started to emerge. Internet telephony is one of them. And like the Internet itself, Internet telephony is pretty easy to do. Today you really don't need any special equipment to start talking over the Internet. If you've got a multimedia PC, microphone, the right software, and a connection to the Internet, you're ready to *speak* with someone over the Net.

I wrote this book with the hope of sharing some of my personal enthusiasm for Internet telephony; when it comes to Internet telephony, I've even been called an evangelist. Some people just call me obsessed. But like it or not Internet telphony has already rattled the cages of the global telecommunications industry and might just be the most significant way to ultimately reduce the cost of your long distance phone calls.

Chapters 1 through 4 provide the background and history of Internet telephony. I also include an overview of some of the common problems you're likely to experience when first using some of the products. In Chapter 4, I discuss some of my predictions for future uses of Internet telephony. Along the way you'll also find information pertaining to my own personal history with these enabling technologies.

Chapters 5 through 12 focus on the following products and how to use them:

- Internet Phone
- WebPhone
- WebTalk
- TeleVox
- CoolTalk
- FreeTel
- NetMeeting

Chapter 13 talks about other products like VDOPhone, PowWow, Stream-Works, and many more.

This book is for you if you are new to Internet telephony or perhaps the Internet and want to learn more about the concepts that you've heard a lot about in magazines and in the news. This book is also for those of you who want to use the products described in this book to make free long distance phone calls within your own personal calling circles. And yes, if you're involved in business and want to use the Internet to lower the long distance costs between you and your business partners, read on.

I was one of the early users of Internet telephony software products and along the way I've tested and evaluated each of the various Internet Telephone software products that have become available since February 1995.

Within a week of the introduction of the first wave of Internet Telephone software products, I set up several mailing lists where users of these products could get together to discuss the merits of them, and I also established several Web pages dedicated to this and other emerging and enabling technologies. Please feel free to visit http://www.pulver.com to catch up on all of my latest interests and activities.

With this book you will gain the insider's track to who the various players are in this marketplace and how you and your business partners, friends, and family members can jointly start talking on the Internet, start having some fun and save a significant amount of money on your long distance phone bills.

JEFF PULVER

jeff@pulver.com

August 1996

Chapter 1

The Concept:
Free Long
Distance Calling!

How would you like to make long distance phone calls for free? Well, you can. Not only is it perfectly legal, you can start doing so today. How? By using the Internet. Yes, you can use the same "Net" you use to view home pages on the World Wide Web and send e-mails to reach out and talk with people from all over the world.

Imagine never having to worry about how much time you spend on the phone making long distance calls—speaking with your high school friend who moved to the West Coast or with your business partner located on the other side of the world. In many cases you can spend an unlimited amount of time speaking with people without having to pay any more than the incremental cost of your Internet connection. The technique of using the Internet as a telephone is commonly referred to as Internet telephony, or sometimes voice on the Net (VON).

How Internet Telephony Works

To use voice communication over the Net, you'll need a microphone and speakers in addition to your personal computer. When you speak into your microphone, your voice is turned into digital packets that are routed over the Internet to the person with whom you are engaged in a conversation. When two people use Internet telephony software, they communicate by speaking in their normal voices, the sound of which is converted into digital information, called packets. When your digital packets arrive at the receiving side, they are converted from digital representations back to the sound of your voice, and you become audible through the receiving party's speakers.

All of this happens in "real time," so your ability to speak with somebody else can be affected by many things, but it all goes on behind the scenes. All you need is Internet connectivity and the right software; the Internet takes care of the rest.

Some Internet telephony software products such as Internet Phone allow you to monitor statistics of your calls. These statistics contain information about the average delay between the two parties as well as the packet loss. On the Internet, it is common for the average delay during a conversation to be between a half a second and a second. If you run into a problem while you're speaking on the Net—for instance, if, without warning, it becomes difficult to understand the other person—you could be experiencing packet loss. Most Internet telephony software products allow conversation to occur when the overall packet loss is less than 10 percent. When the packet loss is greater than 10 percent, words start sounding very choppy and entire sentences may start to drop out. Usually, though, the problem goes away in a few seconds.

Who Uses Internet Telephony?

Now that I've got your attention, let's talk about who is actually using this technology.

Today, there are three main groups of people who are speaking over the Internet:

- Hobbyists who have discovered a new form of global communication

- Individuals who are using the Internet to speak with people from their "personal calling circles"

- Business professionals using the Net to speak with people in their office or in the offices of individuals they do business with

There might be a fourth group of Internet telephony users:

- Telecom executives who are trying to assess the risk Internet telephony poses to their business.

In addition, Internet telephony is used by many colleges and universities that offer distance learning classes. Although English is not spoken by everyone in the world, many of the international users of Internet telephony speak English, so there is a good chance for common understanding. If you know a second or third language, Internet telephony offers a great opportunity to practice your language skills.

The size of the Net population that has embraced this exciting new technology has been growing rapidly, with members from the hobbyist, individual, and business communities. During 1995 the majority of users were hobbyists; one research and consulting firm, International Data Corporation, estimated that 20 thousand people a week used this technology. As each month passes, the mixture of Internet telephony users keeps changing. Each day there are people who introduce this exciting technology to their friends and family members as well as their business partners.

In many ways, the Net has become a social community unto itself. As people from around the world discover the Internet, and specifically the appeal of Internet telephony, you'll undoubtedly find yourself staying in touch with people with whom you haven't talked in years, as well as total strangers who live on other continents and in other societies. What you can learn from them is boundless, once the communication barrier has been broken. When Internet Phone was introduced, I remember speaking to sheep farmers in Argentina, taking part in a virtual birthday party in Toronto, and being a caller to a TV talk show in Tel Aviv, as well as a caller to a morning radio show in Detroit. Imagine waking up in the morning, turning on your PC, and spending 15 minutes speaking with a friend of yours in Australia while you are both waiting for your mutual friend from Indonesia to join in.

Thousands of people have been discovering Internet telephony each week, so a wide range of people are available to talk 24 hours a day, 7 days a week. If you're ever lonely or looking for people with mutual interests, you can find somebody on the Net. These days, in addition to viewing somebody's home page or sending an e-mail to them, you can now have two-way conversations, whether those people are on the other side of the world or ten minutes from your home.

These days, whenever I travel and carry my laptop PC with me, I'm able to speak to my wife from any locations for the cost of a local phone call. All I need to do is turn on my portable PC, connect to the Internet, and call home to speak with my wife. Over time, calls like these could save me thousands of dollars a year.

These days, many college students have Net access from their dorm rooms. Many parents have taken to the Internet, not only to reply to their kids' e-mail requests for money, but now as part of a weekly and sometimes daily routine of speaking voice to voice with their children who are far away. Thousands of college students use the Net to call their friends—dorm room to dorm room and university campus to campus—over the Internet as well.

Last summer I had an e-mail correspondence with somebody who needed help getting his sound card working in an old portable computer. Eventually, I learned that the reason for setting up the computer was that his daughter was planning to spend six weeks traveling throughout Europe. With her portable computer and a listing of Western European Internet access providers, the daughter could speak to her concerned parents twice a week instead of sending postcards that would probably arrive home about the same time she did.

Take another example. A Net friend of mine who is in the import/export business was one of the early adopters of Internet telephony. It took him only three weeks to convince his business partners in Hong Kong and London to try out Internet voice communication. These days he is saving more than $2,000 a month on long distance phone calls.

As businesses and individuals learn about these exciting new technologies, which are very easy to use, many more people will begin to start saving money as well.

The Origins of Internet Telephony

People have been talking on the Net since about February 1995 when a small Israeli company, VocalTec Inc., introduced its Internet Phone (IPhone) product. The Internet hasn't been the same since. Table 1.1 shows a simple timeline for Internet telephony products. Internet Phone was the first of many "critical-mass" Internet telephony products employed by Windows users. Today versions of Internet Phone are available for Windows 3.1, Windows 95, Windows NT, and the Macintosh. As new operating systems are developed and gain popularity, it is certain there will be accompanying software that supports Internet telephony in one form or another.

TABLE 1.1 Recent History of Internet Telephony and Other VON Products

Release Date	Product Name	Home Page/Web Address
Late 1992	VAT	http://wwwnrg.ee.lbl.gov/vat/
Mid-1993	CU-SeeMe (Mac)	http://cu-seeme.cornell.edu
Early 1994	CU-SeeMe (Win)	http://cu-seeme.cornell.edu
Late 1994	NetPhone	http://www.emagic.com
February 1995	Internet Phone	http://www.vocaltec.com
April 1995	RealAudio	http://www.realaudio.com
June 1995	PowWow	http://www.tribal.com
June 1995	Enhanced CU-SeeMe	http://www.wpine.com
July 1995	TrueSpeech	http://www.dspg.com
July 1995	Voxware	http://www.voxware.com
August 1995	Streamworks	http://www.xingtech.com
September 1995	Cyberphone	http://magenta.com/cyberphone
September 1995	DigiPhone	http://www.planeteers.com
September 1995	WebPhone	http://www.netspeech.com
September 1995	Internet Wave	http://www.vocaltec.com
October 1995	WebTalk	http://www.qdeck.com
November 1995	VDOLive	http://www.vdo.net
December 1995	FreeVue	http://www.freevue.com
January 1996	FreeTel	http://www.freetel.com
January 1996	PGPhone	http://www..mit.edu/networkpgphone
February 1996	TeleVox	http://www.voxware.com
February 1996	Free World Dial-Up	http://www.pulver.com/fwd
March 1996	CoolTalk w/Netscape	http://www.netscape.com
April 1996	VDOPhone	http://www.vdo.net/vdophone
June 1996	ICPhone	http://www.ibm.com/internet/icphone
June 1996	NetMeeting	http://www.microsoft.com

When Internet Phone was introduced, it quickly became the market leader among those on the Internet who used Windows 3.11 and Windows for Workgroups. For a short time, it was the only "mass-market" commercial Windows product available for Internet telephony. The release of Internet Phone created a lot of media hype and interest. The thought of speaking to people all around the world for the price of a local phone call sounded almost too good to be true. Was this really possible? Did it really work? The answer to both questions was "yes." In the beginning the sound quality wasn't that great and there was always something that didn't work perfectly, yet for the cost of it (free!), the quality was good enough.

While I was writing this book, I was also a columnist for *Boardwatch Magazine.* My first article regarding VON technologies covered the history of Internet telephony products. In response to my history "lesson," I received the following e-mail:

From: Will Price <wprice@primenet.com>

To: jeff@pulver.com

Subject: VON

Hi Jeff!

As the author of PGPfone, I just wanted to note a couple items which occurred to me in reading your cool article about IP telephones.

IPhone was not the first Internet phone. VAT (http://www-nrg.ee.lbl .gov/vat/) was the first Internet telephone, and was the basis for the design of the RTP protocol that has now become the standard endorsed by Netscape, many others, and my own group developing PGPfone. VAT has been available for years. NetPhone for the Macintosh, which was also available before IPhone around 1/95, was based on and compatible with VAT. Looking at your table, I guess you must be referring to Windows products only in which case you're correct.

The IP phone of the future (6 months to a year) will be based on RTP, the VoxWare coder, and the good ones will have the PGPfone protocol for secure calls. The promise of free, secure calls is not far off.

One important note is that RTP is not a complete solution. There is a massive hole in RTP which will continue to make IP phones incompatible. It specifies how to send the audio, but not how to negotiate or begin a call. We'll have to wait and see what happens on that front.

Cheers.

— Will

I sent the following e-mail in reply:

Will,

The history referred to in the chart from the April 1996 article was a reflection as you suspected of the history of Windows IP telephony products.

With respect to NetPhone for the Macintosh—I'm sure that 1995 was a lot of fun for Andrew Green and his team at Electric Magic Company in San Francisco. It just goes to show that a few people with a good idea and connectivity to the Net really *can* make a difference.

— Jeff Pulver

In other words, Internet telephony has been in the works for quite some time.

Netscape Steps In

In early 1996, Netscape Corporation announced its acquisition of Insoft. The forthcoming version of the company's browser, Netscape 3.0, will contain components of something known as LiveMedia, which among other things will provide Internet telephony support within the Netscape browser. Netscape is also well positioned to become the driving force behind standards within the Internet Telephony industry. The Netscape approach is simple: integrate telephony functionality within its browser and see what happens when 30 million people start talking with each other.

LiveMedia from Netscape is an open framework for building real-time audio/video applications for the Internet. Netscape originally used Insoft's technology to create the Netscape LiveMedia framework. When LiveMedia was announced, eleven companies, including Progressive Networks, Adobe Systems, Digital Equipment Corporation, Macromedia, Netspeak, OnLive!, Percept, Silicon Graphics Inc., VDOnet, VocalTec, and Xing, expressed support for the technology.

The Netscape LiveMedia framework is based on the Internet Real-Time Transport Protocol (RTP), RFC number 1889, and other open audio and video standards such as MPEG, H.261, and GSM, which enable products from various companies to work together seamlessly, providing users with a range of real-time audio and video capabilities on the Internet. Netscape has published the LiveMedia framework on the Internet and has openly licensed key technology components of it. Netscape continues to work with the Internet standards bodies (IETF) to facilitate the adoption of this technology as a formal Internet standard.

Easy to Use and Painless to Install

I suppose the number-one reason Internet Phone was successful right from the start was the fact that the product actually worked. The installation took less than three minutes, and after installation, it was easy to establish a voice contact without needing the assistance of the help files.

In the software industry, most companies don't announce and ship the new product immediately, but VocalTec did just that. Furthermore, because the company's initial distribution vehicle was the Internet, people from around the world downloaded the software literally overnight. Within the first 60 days of Internet Phone's availability, people in more than 40 countries were able to talk to each other over the Net.

As far as the initial product acceptance went, it didn't hurt that for those people on the Internet at the time, the installation and use of the IPhone didn't require the reading of any documentation. The most difficult part I encountered was locating the mixer settings program that came with my sound card and to find the right icon to click on in order to enable my microphone. VocalTec also provided a rather comprehensive online help file in case a user did have problems.

About the time Internet Phone was introduced, I set up and established the Internet Phone mailing list on the Net. During the first few weeks of the mailing list's existence, people started posting questions and getting answers. A user community had sprung up. At times it has been hard work to keep up with the demand, but the knowledge transfers that take place benefit the entire readership of the list. Within a day of the list announcement, more than 125 people had signed up. By June 1996, more than 4,000 people were subscribing to the list to keep up with the trends and issues surrounding Internet telephony.

Perfect Timing

One of the other major reasons for Internet Phone's quick success, and for the acceptance of Internet telephony products in general, was the timing of the release. By the middle of 1994, most of the 486 computers being sold were packaged with multimedia kits, so in addition to the standard keyboard, CPU, monitor, and mouse, they also had speakers and a sound card. The only other accessory needed for Internet telephony was a microphone. As 1995 moved forward and the Pentium computers became more powerful, the price/performance ratio became even more evident. As these Internet telephony-ready computers were purchased and used by individuals to connect to the Internet, people had all of the hardware needed to start speaking with others around the world. VocalTec then provided the software.

As more and more people downloaded Internet Phone from the Web, word spread about the concept of free long distance calling. Pretty soon, reporters from magazines, newspapers, and radio were checking out this technology. One of the things I found particularly exciting was because of the interest generated by this technology, I now had easy access to local reporters from New York City. I remember speaking over the Internet with both Dave Brode of WNBC-TV and Fred Fishkin of WCBS Newsradio 88 on several occasions using this technology. The reaction was the same across the board: This really is an exciting new way to meet people.

My first Internet Phone contact was with Andrew Kantor of *Internet World*, but like many others on that first day of operation in February 1995, I didn't have a microphone, so all I was able to do was listen to Andrew speak to me. All I heard him say was, "Hello. Are you there? I can't hear you. Hello? Hello?" Even after I got my own microphone there were still plenty of others who didn't have one, so this trend repeated itself during the first couple of weeks of Internet Phone. In fact, over a year and a half later, I still end up connecting with people who don't have microphones. When this happens, some software products allow you to simultaneously type messages to the person with whom you're trying to communicate.

When I think back on what a kick those first days using Internet telephony were and what got me really "hooked" on the technology, I remember clearly the instant friendships that were formed and the helpful way people treated each other. For instance, when one person was having a problem using a certain feature, someone would be around to help out and lend a hand.

After reading about my excitement regarding the use of Internet Phone, you might think that I am "Internet Phone" biased. You have to remember that, when Internet Phone was released, it really was the only game in town; I became a fan because it actually worked. However, what I was a fan of wasn't so much Internet Phone but the broader concept of talking to people over the Net. Since the summer of 1995, I've had the pleasure of trying out and evaluating more than 20 other products that offer different levels of Internet telephony functionality.

The Internet Telephony "Sound"

Some people have described Internet telephony as being similar to the early days of global satellite communication. My own experiences have led me to compare it in sound quality to cellular phone calls.

Depending upon how well the people you speak with configured their PCs and the kind of microphones they use—together with the quality of the connection both parties have to the Net—the sound quality can be anything from good "AM radio" quality to CD quality, or anywhere in between. If there is high packet loss between the two parties or if there is something wrong with one or both configurations, the sound quality will be poor. However, when you consider that these calls are being made for free, except for the cost of the Internet connectivity on both sides, there should be a little leeway as to the sound quality you're willing to tolerate.

Differences between Internet Telephony and CB/Ham Radio

If you have used CB radio and are used to saying "breaker, breaker" (or, on ham radio, "CQ, CQ") when seeking a contact with someone, one of the things you'll notice with Internet telephony is that there isn't a band to tune into. Instead, depending upon the product you use, you'll see a list of individuals who are "on the air" and who are available to speak. One of the best parts about CB and ham radio is that you can just listen to the conversations that are going on between other people without acknowledging that you're there. You can't do this with Internet telephony. There is currently no way to listen in to others who are engaged in a conversation.

Note. *If you enjoy the ability to listen in without becoming an active participant, I suggest you take a look at the Enhanced CU-SeeMe product from White Pine Software. A copy of it is included on the CD-ROM that accompanies this book. Enhanced CU-SeeMe is White Pine's desktop video conferencing software for real-time person-to-person or group conferencing. The key to this is that the "one-to-many" effect is achieved by using what is known as a reflector. You can set up an Enhanced CU-SeeMe client, point to a reflector, and listen to the conversations taking place among the visible individuals or groups who are also using the reflector.*

Another difference between ham and CB radio and Internet telephony is that, unlike CB where you have "channels" and ham radio where there are specific frequencies on which you can transmit, there really isn't any "Internet band." However, you can think of each Internet telephony product as representing a separate band. Also, unlike ham radio where you can find people using different modes of communication—whether it's Morse code, single-side band (SSB), or slow-scan TV—with Internet telephony only one mode is currently supported. Although the Internet isn't directly affected by sun-spot cycles, propagation, or solar flares, if one of the major Internet routers goes down, it has an effect equivalent to a solar flare on radio—an entire region of the world can and will lose communications, from what amounts to be a moment to as long as several days.

Like everything else, Internet telephony has its place, but it won't replace ham radio anytime soon. For example, while driving their cars, ham- and CB-radio operators are able to listen to their two-meter sets and as a result steer away from tornadoes and other bad-weather situations. Low-cost, two-way wireless communication is getting there, but it's not there yet. When it does arrive, it might take some time to achieve critical mass.

Nickname = Callsign
With the various Internet telephony products, when you're prompted for a nickname, former and current ham radio operators use their callsign as a means of identification.

Internet Relay Chat (IRC) Servers

For those of you who've already been on the Internet, you may be familiar with one of the most popular forms of real-time Net communication: the ability to log into one of thousands of Internet relay chat (IRC) servers and connect with chat sessions taking place among a rather anonymous group of users. Because Internet Phone was designed to act like an IRC, it can make use of the global IRC server network. Sounds good, right? Well if you were one of the IRC server operators (IRCOPs) back in March 1995 you may not have felt the same way. The IRCOPs felt that the use of Internet Phone—especially the original trial version that at the time allowed only 60 seconds of chat time before the need to exit and reconnect to the server kicked in—was consuming resources that were already strained to begin with. There was a concern that continued usage might cause some operational problems with the entire global IRC server network.

As a preemptive strike, the people who ran EFNet—the organization responsible for the IRCOPs around the United States and parts of the world—decided to put into their network a patch that basically banned Internet Phone use over the Internet.

While the EFNet went through the motions of having Internet Phone banned from the membership of the global EFNet, I was on the Net monitoring the situation. Talk about being at the right place at the right time! For reasons that I still have trouble explaining to even myself, I took the initiative of setting up a private IRC server on the dedicated 56K connection to the Net that I had at my home. When I announced to the Internet Phone mailing list members that I had set up a separate Internet Phone IRC server, people started to drop by and check it out.

It took about a week working with VocalTec, but within that short period of time, a private IRC server network was born. Soon there were servers running in the Netherlands, in Washington, D.C., and on Long Island, New York.

During times of national and international disasters—whether the Kyoto earthquake or the bombing of the federal building in Oklahoma City—these real-time chat services are one place on the Internet with the best coverage and up-to-date information. Although the people involved many not be professional reporters, chances are you will find somebody with Net connectivity who is willing and able to provide information pertaining to a particular event.

Although it may appear that nobody "runs" the Net, groups of relatively unknown people can have a profound effect on others. Witness what happened when Internet Phone was introduced. Recognizing the benefit of the worldwide IRC server network, Internet Phone was designed to be a "client" that connected to the IRC network. This seemed to make sense to the developers of Internet Phone, who wanted to develop a software product that could connect to an existing public network of users, with thousands of local servers and a place that, at any given moment, thousands of people could be online and available for conversations.

Directory Services

I believe that because of the "IRC server war" described above, some Internet telephony software companies made it a point to include in their marketing materials a statement that they don't use IRC server technology to allow people to connect with each other. Frankly, I think those companies are missing the point. The IRC server simply provides what is known as "directory services" to the person using the product. From a technical view point, these products don't actually connect to each other via the IRC server. The IRC server merely provides a way for you to see who else is around. Once you decide who you want to communicate with, your conversations take place directly with the other person using what is known as direct UDP/IP connection.

Directory services are dedicated servers with which an Internet telephony product connects so users can determine who is online. FreeTel's server even tracks who is engaged in a conversation and who is available. The operation of these directory servers are sometimes also referred to in general, collective terms as a server-based model. Internet relay chat servers on which VocalTec relied for Internet Phone users to find each other are another example. These servers also serve the purpose of sometimes associating an e-mail address with an IP Address. This server-based model provides the enabling technology that allows two people to connect to the server and establish a direct conversation with each other. Once the Internet telephony industry converges and commercial product standards are agreed upon, the issue of having a shared common directory will also need to be addressed.

Of the Internet telephony products that were introduced after Internet Phone, many have the ability to initiate a direct call with somebody—without the need for the user to first locate the person in a specific directory on a specific server. Whether or not this feature is really needed, it provided a perceived competitive edge for those companies that implemented it. It looks as though VocalTec liked the idea as well—even Internet Phone 4.0 supports direct calling.

A server providing directory services generally allows users to create "topics" or "rooms" that people enter, or join, for the specific purpose of finding others who share the same interest. Some servers support the creation of private topics, which—although they are not password protected—you can join if you know the name of the topic.

IP Addresses

There are generally two types of IP addresses on the Internet: static and dynamic. If you have a dedicated connection to the Internet, the IP address of your host machine does not change. This is an example of a static address. On the other hand, if you have a dial-up SLIP/PPP Internet account, the IP addresses you are assigned, in most cases, will change each time you connect. This is an example of a dynamic address. This means that while some IP addresses are unchanging, many people who access the Internet are assigned a different IP address each time. Even though e-mail will always go to a pre-selected destination (read "static") that is stored at your Internet service provider until you log-in to retrieve it, if you want to establish an Internet telephony contact with somebody who doesn't use a permanently assigned IP address, the server providing the directory services will resolve e-mail addresses into IP addresses.

In order for two people to speak to each over the Internet, the software products they use need to know the exact IP address of the other. Internet telephony software products generally use two types of connections:

- A TCP connection from the individual workstation to the directory server

- A UDP point-to-point connection between each of the parties involved in the conversation

Other Internet Telephony Product Features

Two features that are often found in Internet telephony products include the following:

- *Text chatting service.* These services provide you with the ability to type messages on your screen to two or more people at the same time. If you have ever used the IRC before or tried out the "CB simulator" on CompuServe, you're familiar with the concept of text chatting. However, one difference is that the Internet telephony service is generally private, and there will be a limited number of people who participate with you. In the event someone you want to communicate with doesn't have a microphone or is having trouble with their sound card, you may find yourself using the text chatting service instead of speaking with someone using your voice.

- *File transfer mode.* This feature allows you to transfer a data file while you are in communication with another person. Even though some of the other Internet telephony features allow you to communicate in what is sometimes referred to as a "one-for-many" situation, when you transfer files it is generally when you speaking with just one other person.

One of the eventual improvements in the original Internet Phone product was the implementation of additional voice compression technologies known as a codec. The codec is responsible for the compression/decompression of the audio information and into the digital environment.

In June 1995 when Internet Phone 3.0 shipped, it was the first publicly available Internet telephony product that included support for full-duplex communications. Full duplex is a feature that allows you to speak at the same time as the person you are talking to, somewhat like a live, face-to-face or phone conversation. Full duplex differs from half duplex in that the latter has the same sound quality as a speakerphone. In other words, you can't hear the other person speaking while you're talking.

What Do the Phone Companies Say about Internet Telephony?

At first, most of the major telcos simply ignored Internet telephony. Most companies did not assess the product as a short-term risk.

That changed in March 1996, when a trade association known as ACTA—American Carriers Telecommunications Industry, a group representing 130 lesser-known long distance telephone companies—got together and petitioned the Federal Communications Commission to ban the sale and use of Internet telephony.

It is hoped that this action will not be taken, but it shows that the phone companies are no longer discounting the real opportunities of Internet telephony and related technologies.

Chapter 2

Getting Started

"So . . . how do I get started calling on the Internet and lower my long distance telephone costs?"

Where to Begin

If you already have Internet access and a multimedia PC, you're very close to getting started using Internet telephony products. And if you're not on the Internet just yet—that's cool, too. In order to be "Internet-phone ready," you will need to take a few extra steps.

A multimedia PC in this case is a PC—486/25, or faster, that has a sound card and hopefully a CD-ROM drive. The Internet telephony software products described in this book require only the sound card. But if you want to use the CD-ROM that is at the back of this book, you will need to have a CD-ROM drive as well.

I believe you will find that making long distance calls using the Internet can be easy, assuming you have all of the required tools. In fact, you might start feeling a little guilty about it or start worrying that somebody is watching over your shoulder, because you have found a way to bypass your local telephone company. Don't worry about it. Using the Internet to make phone calls—whether they are local or long distance—is perfectly acceptable, and after a while you will begin to take it for granted.

This is not another one of those AT&T commercials that says something like "It will happen, and we will be the ones to bring it to you." The reality of Internet telephony is that with the proper setup and the right software, you have the ability to communicate with anybody in the world—today—as long as you and the person you're communicating with have Internet access.

If you have ever used a ham or CB radio and you enjoy chatting with others, you'll like Internet telephony. At any given moment, literally thousands of people are waiting to talk to you. The culture surrounding Internet telephony is slowly evolving each day, and by going online you can and will become part of it. If you are in the mood for lying back and talking with just about anybody, you've come to the right place. In fact, there's already been one documented marriage that resulted from two people meeting on the Internet and speaking with each other.

In addition to people who like to chat with each other, there are thousands of people who use the Net every day to keep in touch with friends and family members all over the United States and from around the world. I have also started to see various businesses setting up links to the Net from their corporate websites. They are able to accept incoming calls from the Internet and route them to the appropriate people within their office.

Whether you plan to arrange a talk schedule with somebody you know or are planning on just installing the software and speaking with the first person whose name appears onto your screen, you can do either. No need to worry or feel nervous about talking with strangers. If you've ever found yourself speaking with somebody who you didn't want to talk to on the telephone, or if you've ever wanted to get off the phone but just couldn't, here is a secret that works with Internet telephony just as it does in real life: just hang up. The other person won't suspect a thing. If you find yourself speaking with somebody and you are not enjoying the conversation, there's no need to continue. When it is your turn to talk, find the "hang up" or Exit button on the software product and remove yourself from the conversation. Life is just too short to find yourself spending time speaking with people whose conversations you aren't enjoying.

One of the coolest things I have noticed about talking on the Internet is that it is one of the places where you can go and everybody really *does* know your name. At times I feel as though I'm walking into a local bar and sitting down and speaking with the people around me. After you've been using any of the products for a period of time, some of the people will become familiar to you. No matter where you "travel" around the world or what time of day or night it is when you decide to go "on the air," chances are you will find one of your telephony friends. By the way, on any given night, one of the people you might find "on the air" is me. I usually go by the Nickname of Jeff62. If or when you find me online, feel free to drop by and say hello.

No Brand Loyalty

Unlike people shopping in the supermarket for their favorite cereal, I have noticed that some people, even those who shelled out the cash to become registered users of a particular product, are very fast to jump to another product if that particular product is perceived to have more features and is more reliable. I'm no different. Every time I read about a new product announcement from the voice on the Net (VON) mailing list (see Appendix A) or one that is listed on the Internet telephony resource page on the Net (see http://www.pulver.com), I have a compelling urge to download the software and try it out. If the product really works as indicated in the press release, it usually becomes my Internet telphony product of the moment. For this reason, we've supplied a lot of these products on the accompanying CD for you to play with and find your favorite.

Use the Right Equipment

The real catch is to get set up with the right equipment the first time around. This will reduce your overall frustration level and give you a really good chance to achieve immediate success the first time you try these products. In terms of Internet telephony, this generally means you should purchase the following pieces of equipment to improve your conversation:

- A high-quality microphone that is balanced for the sound card that you use

- A 28.8K modem

Really Getting Started

In order to get started using Internet telephony (and start having fun), you will need a few key items. One of them is a multimedia PC with Internet access and a speaker and microphone. If you don't have Internet access yet, or don't have a sound card for your PC, now would be a great time to go out to a local computer store and purchase a sound card and a modem.

What Are the General System Requirements?

To get started, you'll need the following pieces of basic equipment:

- *Computer:* IBM-compatible PC, 486/25 or greater; at least 8MB of RAM.

- *Internet connectivity:* You will need to run products that require the use of TCP/IP. TCP/IP stands for: Transmission Control Protocol/Internet Protocol. Generally speaking, this means that you will have either a dial-up SLIP/PPP account or a direct connection from the Net—from your dorm room, office, home, or wherever you go to get "connected."

Note. *If you are accessing the Net from your office and you are behind a firewall—unless the firewall has been modified to allow the TCP and UDP packets from your Internet telephony product to travel in and out of the firewall—you may not be able to use the software product successfully. In this case, contact your internal network security group and have them speak with the technical support team of the product you are planning to introduce into your firm. Checkpoint is one of the first Firewall developers to add specific support for some of the more popular Internet telephony software products.*

If you don't have Internet Access yet, you need to set up an account with one of the many Internet service providers (ISP). Your account will probably be a dial-up PPP or SLIP connection, and in the United States you can usually get connected for $20 to $30 per month. Your dial-up connection should be at least 14.4K baud (minimum); most of the time you will experience improved sound quality on a 28.8K connection.

If your Internet service provider is one of the U.S. national companies such as CompuServe, America Online (AOL), Earthlink, Microsoft Network, Netcom, GNN, or PSINet, all you need to do is dial out to your local access number and then fire up the underlying software. All of these vendors (and others in the near future) provide you with SLIP/PPP access, which means that your computer will be able to run Winsock applications. A Winsock application is a windows-based program that uses TCP/IP and relies on the Winsock.DLL. Most of the local Internet service providers can provide you with SLIP/PPP access. If you have any doubts regarding whether or not you have SLIP/PPP access, contact your ISP's customer service desk.

If you can run third-party network applications (such as Netscape or FTP clients) not created by your service provider, then you should be able to run Winsock applications.

When you access the Internet, you need to make sure that you are using TCP/IP. If you are using an ISP to connect to the Internet, chances are that you will be connecting using something known as SLIP or PPP. If you are directly connected to the Net from your dorm room or university office, chances are you would be connected with an ethernet connection via a network card in your PC.

If you are not able to establish a SLIP, PPP, or direct connection to the Internet, chances are that your PC isn't running what is known as the Winsock.DLL. The popular Internet telephony products all require that the Winsock.DLL is loaded prior to loading the telephony product into memory.

If you don't have your own e-mail account, you should arrange to set one up. Whenever you become a subscriber to an ISP, you can get an e-mail account assigned to you. There are many advantages to having an e-mail account. Furthermore, you need one in order to use some of the products discussed in this book. Some Internet telephony products require your e-mail address in order to register their software products.

Another good reason to have an e-mail account is so that you can set up a schedule with a friend when you want to arrange a later meeting or conversation. Some of the VON products actually require that you use an e-mail address as your "phone number."

If you have your own e-mail account, you can also subscribe to a mailing list for people who are interested in leading-edge technology. To subscribe to the list, called the VON Mailing List, send an e-mail request to majordomo@pulver.com. Leave the subject blank; in the body of the message, write "Subscribe von-digest." On the CD-ROM that accompanies this book, I've included a copy of the entire 1995 and part of the 1996 versions of the mailing list.

One word of caution: if you have a UNIX Shell account that you use to access the Net, you will need to have your system administrator install TIA in order for you to be able to run Winsock applications.

Other pieces of equipment you will need include the following:

- *Operating system:* Windows 3.1, 3.11, Windows for Workgroups 3.11, or Windows 95. The Internet products discussed in this book should also work with Windows NT.

- *Sound cards (8-bit or 16-bit):* Creative Labs' SoundBlaster 16 and the AWE32 are currently the most popular sound cards (see Figure 2.1). Both the SoundBlaster and the AWE32 are capable of operating in full-duplex mode. A number of other sound cards on the market provide half-duplex but not full-duplex ability. If you plan to run other software besides Internet telephony-related packages on your PC, make sure that the sound card you purchase is compatible with SoundBlaster.

- *Optional CD-ROM:* This will be a useful accessory if you would like to access the programs and information contained on the CD-ROM that accompanies this book.

Half Duplex and Full Duplex—What's the Difference?

In theory, full duplex transmission allows you to talk to somebody at the same time as they are talking to you. Half-duplex mode allows only one party to speak at a time. In reality, due to the slight delays associated with speaking with someone using Internet telephony, many people use half-duplex mode even though their hardware supports full duplexing.

FIGURE 2.1 One of the more popular sounds cards from Creative Labs.

If there ever was a top-ten list of overhyped features dealing with Internet telephony software products, full duplexing would be at the top of the list. Full duplexing does sound like a great selling feature compared to speaking and then listening and then speaking—which is the same kind of feeling you might get when using a speakerphone. However, on the Internet, you will experience delays from 400 milliseconds to more than a second when you are speaking to someone, so although you *can* both talk at the same time, you really can't take that much advantage of it due to typical network delays.

Tools of the Trade

One of the first things I did after I realized that I had become hooked on Internet telephony was to purchase a set of headphones. Similarly, if you find yourself getting involved in Internet telephony as a hobby or because you want to save yourself or your business some money, you may want to make an investment in some of the items that can have a profound effect on the quality and comfort of your Internet phone calls.

With the arrival and rapid growth of the Internet telephony industry and as more people like yourself decide to try out these exciting technologies, an entire cottage industry of accessory products has started to emerge.

The Internet telephony accessory marketplace can be broken down into the following categories:

- Specialized headsets and handsets
- Microphones
- Speakers
- Sound cards

Specialized Headsets and Handsets

For those of you who want hands-free operation while talking on the Net or want to take advantage of full duplexing, you will need to use either a headset or handset. For full-duplex operation, a headset is really mandatory. You will experience audio feedback between your microphone and speakers otherwise.

Some of the more popular handset and headset products are manufactured by these companies:

- Jabra Corporation
- interActive Corporation
- Firecrest Group
- RadioShack

Jabra Corporation

The Jabra Corporation (http://www.jabra.com) has developed the tiny Ear Phone, which fits in your ear, as shown in Figure 2.2. The Ear Phone allows hands-free, full-duplex communication. Both the microphone *and* the speaker fit in your ear. With the Ear Phone—assuming that you are using a full-duplex sound card—you can talk and listen at the same time.

interActive Corporation

The interActive Corporation has developed a specialized telephone handset that looks like a telephone and can be attached to the side of your computer. If you look at Figure 2.3 quickly, it does look like a telephone.

Firecrest Group

The Firecrest Group resells a product called the Internet Transphone that brings all the convenience and privacy of conventional telephone technology to the Net (see Figure 2.4). The Internet Transphone is really a secure,

FIGURE 2.2 The Jabra Ear Phone.

FIGURE 2.3 The SoundXchange handset.

computer-controlled encryption hardware system that supports the Secure Electronic Transactions (SET) standard developed by Mastercard and Visa. It even has an electronic strip built in so that one day you will be able to slide your credit cards through and order products over the Internet—while speaking over the Net at the same time, using the Transphone handset.

FIGURE 2.4 The Firecrest Group's Internet Transphone.

Radio Shack (Tandy) Corporation

Many people have also been happy using the Radio Shack (Tandy) Pro 50-MX headset More information is available from the Radio Shack home page at http://www.tandy.com.

Microphones

The quality of the sound received by the person you are speaking with is directly related to the quality of the microphone you are using and how much background noise is present in the room when you're talking.

Microphones come in all shapes, sizes, and price ranges. You need to check your sound card documentation for the type of microphone you must use—either high impedance or low impedance. You should find good results from both the cheaper desktop microphones ($10–$18 price range) all the way up to the deluxe models, such as the $350 desktop audio-technical MT-858 from Coherent Technologies.

I have found that good sources for high-quality microphones in the United States are the Radio Shack (Tandy) stores, which can usually be found in most shopping malls. Their line of professional audio microphones—specifically, the unidirectional microphones—have been good performers.

Coherent Technologies—if available in your area—is another good source for high-quality microphones.

If you haven't already done so, connect your microphone to your PC, as shown in Figure 2.5.

Figure 2.6 shows how to open the sound mixer program for SoundBlaster.

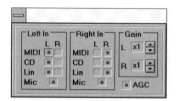

FIGURE 2.5 A microphone that is connected to a sound card.

FIGURE 2.6 The SoundBlaster mixer program.

Figure 2.7 shows the mixer settings.

Make sure that your SoundBlaster software is installed properly and that you can hear your own recording. To do so, turn on the microphone. Usually this is done by using the left mouse button to select it.

Now would be a good time to open the recorder program, as shown in Figure 2.8, and press Record. Speaking clearly into the microphone, say a few sentences. Now listen to your recording. If the digital image of your voice sounds close to normal, then you're ready for your first Internet conversation.

Speakers

I've been very successful using the low-end speakers that came with my multimedia computer. If you are purchasing a set of speakers just for Internet telephony, you really don't need to spend a lot of money in order to get good results.

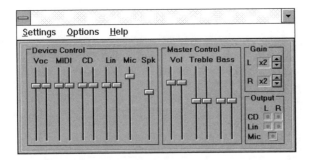

FIGURE 2.7 The SoundBlaster mixer settings.

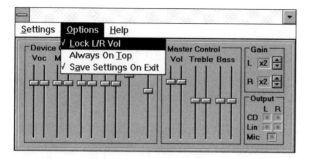

FIGURE 2.8 The SoundBlaster recorder.

Now it's time to check the volume levels/settings for your speakers, as shown in Figure 2.9. Generally speaking, you don't want the microphone to be too close to the speakers, because that generates feedback.

Now use the recorder a few more times, speaking closer to and then further away from the microphone. You will want to get the settings such that you feel most comfortable and most natural for an extended conversation.

By taking the step of adjusting the mixer settings in advance you're avoiding some of the pain associated with trying to figure out the answer to the age-old question, "How come I can hear someone talking, but they can't hear me?"

Sound Cards

There is a wide variety of sound cards available in the marketplace, including those from the following companies:

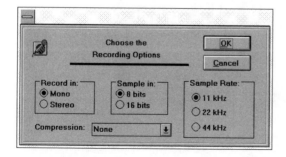

FIGURE 2.9 The speaker settings on the SoundBlaster mixer.

- Creative Labs
- Turtle Beach
- Gravis

Creative Labs

Creative Labs is the maker of the most popular sound cards for Internet telephony. Both the SB16 and AWE32 products are full-duplex ready, and if you already have the sound card but are using it in half-duplex mode, simply download the latest full-duplex driver from the Creative Labs website, as shown in Figure 2.10.

Turtle Beach

The Tropez product from Turtle Beach is a multipurpose, 16-bit sound card that offers digital audio, game compatibility, MIDI synthesis, sampling and recording utilities. For more information on Tropez, check out the Turtle Beach website, shown in Figure 2.11.

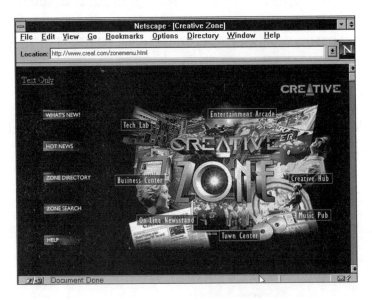

FIGURE 2.10 The Creative Labs home page.

FIGURE 2.11 The Turtle Beach home page.

Gravis

UltraSound from Gravis is a full-duplex sound card (see Figure 2.12 for the Gravis website). Although it is not as popular as the Creative Labs

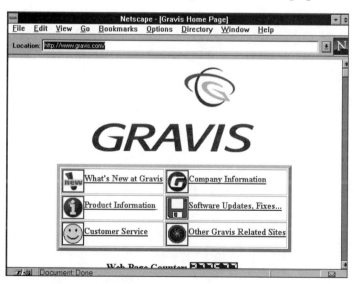

FIGURE 2.12 The Gravis home page.

card, it has been used by thousands of Internet telephony users around the world.

Get on the Net!

If you happen to be the first person on your virtual block to use Internet telephony, the next thing for you to do is visit a Net directory of users of Internet telephony products. One such directory is the Internet Phone Users Directory, which is available at http://www.pulver.com/iphone. There you can find thousands of others who have discovered Internet telephony, and you should have no problem finding someone located close to you geographically, so you can get tips on where to shop for products. Most of the time, people enjoy the opportunity to answer questions and pass along tips to newcomers.

Chapter 3

Typical Problems with Internet Telephony

Before we jump to Part Two of this book and start trying some of these products, we should spend a few minutes going over some Internet telephony basics. When you are ready to go "on the air" for the first time, you should be aware of some problems you might run into.

The Top Ten Problems with Internet Telephony

As David Letterman might say: From the Home Office in New York, here are the top ten things that could go wrong when you use Internet telephony products:

10. *You don't speak the language of the person you just connected to.* This has happened to me on many occasions. Just hang up and find somebody else to speak to.

9. *You called somebody who didn't want to be disturbed.* After the person stops grunting at you, just hang up.

8. *You received an obscene Internet phone call.* Either ask for the person's number so you can follow up, or just ignore it. It's up to you.

7. *You want to talk, but none of your friends are around.* Take the plunge and try calling somebody new. You run the risk of items 10 through 8 above, but it is worth the risk. Internet telephony can be a lot of fun, especially when you speak with people you don't know.

6. *Did you get a look at those X-rated topics?* You just gave a demo of Internet telephony to your boss, and he saw the list of XXX-rated topics—and you just convinced him that the time you spend on the Internet is valuable for the company.

5. *What do you mean, you won't talk to me because I'm not registered?* Some people on the Net like to speak only to people who have made the same level of commitment that they have. Being a "registered" user of a particular product means that you have made the investment and actually purchased the software product.

4. *Where is everybody tonight? Am I the only person online?* Sometimes the Net experiences outages, and it's possible that you won't be able to speak with anybody. This is known as a router outage.

3. *You keep getting disconnected from the servers.* If this happens, try, try again. When the Internet gets busy and a lot of people are online, disconnections are more frequent.

2. *You want to speak with your friend who uses WebTalk, but you are using CoolTalk. Why can't you speak with each other over the Internet?* Until all of the major Internet telephony vendors get together to agree on standards, you will not see two unrelated

products with the ability to communicate with each other. On the positive side, with the advent of Netscape's LiveMedia initiative, it is possible that sooner rather than later we will see interoperablity between vendors.

1. *You can hear the other person talking, but he can't hear you.* Check to see that you remembered to connect your microphone. If you did, check to see if you remembered to open the mixer application that enables your microphone.

In this chapter I've provided answers to some of the hundreds of questions that I get from people who are having trouble with their favorite Internet telephony products. If you have a specific question and you are not able to get a response from the customer support department of your Internet telephony software company, please feel free to send your question(s) to me at jw-questions@pulver.com.

Most Frequently Asked Questions (FAQs)

Here are some of the most frequently asked questions about Internet telephony, as well as their answers:

Q. I can generally connect with other people, but the sound is usually hard to understand. What could cause this?

A. Several things, including the following:

- *The quality of your microphone.* I found it amazing how much difference a good microphone had on the quality of my conversations. At first I was using an old tape recorder microphone I had around the house. When I finally replaced it for a $20 microphone, not only didn't I need to shout anymore, but I also started getting comments from the people I was talking to about how great my audio sounded.

- *The speed of your Internet connection.* Higher-speed connections generally result in better-sounding audio. What happens is this:

The software products are "aware" of the speed of the connection and choose to use different audio compression techniques, based on how much bandwidth is actually available.

- *The amount of traffic on the Internet between the locations of the people who are communicating.* In technical terms, you might want to find out the ping turnaround time and, if possible, the number of hops between the two sites.

- *The speed of your computer.* For example, a Pentium 133MHz will perform better than a typical 486/25. Some products, such as VDOPhone, need at least a Pentium 100MHz chip to perform at an adequate level.

Q. Can I play music from my CD drive and use Internet telephony at the same time?

A. No.

When you play an audio file—whether it's a RealAudio, Streamworks, or .wav file, or from some other source—it will prevent your Internet telephony product from sharing the same resources. When these programs need to gain control of the sound card and it's already in use, you will probably get an error message.

Q. I've heard that there may be better settings for my TCP/IP stack than the ones I'm using. Is this true?

A. The reports I've received seem to indicate that if you are accessing the Internet via a dial-up program, the default settings are generally just fine. Most of the products covered in this book perform well with the Trumpet Winsock stack. For instance, the help file for WebPhone suggests that if you are using WebPhone, you should have the following minimum values defined in the Trumpet Winsock set-up screen:

MTU: 552   TCP RWIN: 2048   TCP MSS: 512

Q. When I'm looking at the technical specifications of sound cards, what information is important to me?

A. The sound cards of today come with many features, most of which will not affect your Internet telephony usage. However, it is important to verify that your cards support both an 8KHz and 11KHz sampling rate.

Q. I'm having trouble configuring the sound levels in this software product. What can I do?

A. Many Windows-compatible sound cards allow you to make sound adjustments using your Internet telephony product. Then again, if you are using an 8-bit sound card like the Creative Labs SoundBlaster 2.0, your ability to make adjustments to the environment is very limited. If you find that you enjoy using Internet telephony, I'd suggest that you consider purchasing a 16-bit sound card in the near future. One of the things I didn't like about my SoundBlaster 2.0 was the fact that it didn't come with a mixer adjustment.

Q. People have suggested that I use AGC. What is it?

A. If you are using at least a 16-bit sound card like the SoundBlaster 16, it is possible to get better audio performance from your card if you enable the automatic gain control (AGC) feature of your sound card. Generally speaking, when you open up the mixer application, you will have an opportunity to enable AGC on your sound card.

Q. What are the TrueSpeech and GSM audio codecs I've heard others talking about?

A. First, the term *codec* stands for "compression/decompression." Global System for Mobile Communications (GSM) is the European standard for digital cellular communications. This provides just about a 5:1 compression ratio of raw audio. On the other hand, some of the technology showcased in TeleVox provides up to a 53:1 compression. TrueSpeech from the DSP Group is another codec somewhere in the middle; it provides a 15:1 compression on 8KHz audio.

Q. Is it really possible to put two sound cards in my PC in order to get full duplex?

A. Well, yes, at least in theory it can be done. I've spoken to several people who claim to have done it. But unless you are extremely familiar with the inner workings of your computer and understand how sound cards work—and if you *must* operate in full duplex mode—I strongly suggest you purchase a sound card (such as the SoundBlaster 16) that is well known among those who use full duplex on a regular basis.

If you already have the SoundBlaster 16 card and you are still operating in half-duplex mode, drop by the Creative Labs website at http://www.creaf.com and follow the links to download the company's full-duplex driver.

Q. My received audio is delayed and choppy. What causes this?

A. This can happen if you are using the Net at a time when there is a lot of traffic. It could also happen if the person you are talking to doesn't have the right kind of equipment or if there is something wrong with his or her connection to the Net. Another possible cause could be the speed of your computer. Some of the newer codecs used by Internet telephony products perform much better on a Pentium class computer.

Try contacting another person. If this problem persists, I suggest you leave the Net for an hour or two and try again.

If you have access to the Finger program, one thing you could do is issue the following Finger command: finger @status.psi.net. PSINet is one of the only Internet Service Providers in the United States that provide an up-to-the-minute status report on not only the state of the network but the state of the MAEast router—the main entry point in the United States for many ISPs. When MAEast is having trouble, which does happen from time to time, everybody who is online at the time feels it.

Q. I can't hear the person I'm supposed to be talking to. What can I do about it?

A. Check the setting of the Internet telephony software product you are using. If you are using a product like WebPhone that indicates the "Talk" mode and you never see the indictor "Listen," it could be a VOX problem.

What this means is that you need to adjust the balance between your microphone activation level and the speaker levels.

I've also run across this situation and discovered that the other party didn't have a microphone. In situations like these, the text chatting software that is available in a number of Internet telephony software packages comes in handy. This is the only way I found out that the person I was in contact with could hear me, but the reason I couldn't hear him was because he didn't have a microphone.

Q. The person I am speaking with is having trouble hearing me. Is there anything I can do about it?

A. If the person you are trying to communicate with doesn't have his or her system properly balanced between the speaker and microphone settings, they could have a VOX problem. If the Internet telephony product you are using has a text chatting feature now would be a good time to use it.

Q. Somebody told me that I might have a "slow" serial port on my computer. What does that mean?

A. If you are using an external modem, you might want to verify that your COM port has the high-speed 16550 UART rather than the older 8250 UART. These days, it is rather inexpensive to install the high-speed 16550 UART. As it turns out, the older 8250 UART doesn't perform well under Windows, especially if you are trying to dial out at speeds faster than 14.4K.

If you want to check out which kind of UART you have, assuming you have DOS 6.0 or greater, run MSD from DOS. When MSD finishes examining your computer, select COM Ports. For each COM port on the system, MSD will show whether it uses the 16550 UART or not.

If the active COM port is Port 2, and assuming your modem is on COM2, make sure that you have the following line in your SYSTEM.INI file, under the [386Enh] section:

COM2FIFO=1

Substitute the number of your COM port with the number 2.

Q. I'm having trouble using full duplex with Internet telephony. Any suggestions?

A. If you are using a dial-up connection to the Net, in order to use full duplex, you need to disable modem compression. Now here's the catch—there is no general way to disable modem compression. It depends on both the modem and the Winsock stack you are using. A good starting point is to check your Winsock documentation and setup to see if a modem compression option exists. If so, follow the directions to disable it. There is a good chance you will need to add an initialization command to disable your modem compression. The initialization string to disable modem compression for most Hayes-compatible modems is AT%C0. This should work with modems from: AT&T, Boca, Zoom, Compaq, Gateway 2000, Intel, NEC, Supra, and Motorola. For US Robotics and Zyxel modems, use AT&K0.

Chapter 4

The Future of Internet Telephony

Sure, everyone would like to look into a crystal ball to get a glimpse of the future. I don't have a crystal ball (yet), but in this chapter I attempt to project some future trends regarding Internet telephony.

Let's look back at the recent history of Internet telephony and see where things have been. Doing so should give us a good indication of where things might be heading in the immediate future. As I mentioned in Chapter 1, hobbyists have been the primary users of this technology to date. Along the way, some executives also shared a vision of potential savings for their companies and took the steps needed in order to prepare for or implement this technology.

Use of Internet Telephony Within Corporate Calling Circles and Networks

Dennis Minero of Long Island, New York, was one of these early visionaries. Dennis used to incur over $2,000 a month in long distance calls. Today, he's reduced his long distance costs to less than $200 a month. He accomplished this by convincing his business associates to converse using Internet telephony rather than a traditional telephone. Many other companies have also made the switch and prospered from the savings. As word spreads about this technology and firms gain Internet access, we will start to see circles following a manufacturer/supplier model.

In the coming months you will hear more stories about how small (and not-so-small) businesses started saving a significant amount of money by using the Internet rather than picking up the telephone and dialing a long distance number.

One of the future applications of this technology is use within internal corporate networks, called *intranets* or internal websites. As companies around the world continue to invest in their network infrastructure and provide end-to-end connectivity between remote offices from around the world, many of the products covered in this book will be deployed in such environments.

For example, let's say that you work for a major brokerage firm in New York City. Let's suppose the company's network runs through 18 cities around the world, including New York, London, and Tokyo. Instead of making long distance phone calls from office to office, the brokerage firm runs an Internet telephony product over the company network. Long distance calls between the various global offices are conducted via the company network.

Another marketplace in which Internet telephony is starting to play an important role is in the small office/home office (SOHO) community. The projected growth of SOHOs and the high usage of computers in this audience warrant the attention this market segment has started to receive from vendors and the press. As SOHOs join trade associations, you can expect these groups to gravitate to technologies that allow them to save money and communicate with others within their community.

One of the trends that I've started to see and believe will continue for some time to come is people and organizations that put their Internet telephony "addresses" on their websites and encourage their customers to connect with them over the Net instead of contacting them via a conventional telephone. Check out the website http://bio.net/aae for an example of one company that has chosen to go down this path. One advantage to using this kind of technology is that the person on the other end of the phone line doesn't actually have to be in the office. As long as the address can be resolved over the net, the person on the receiving end of the phone call can be working from home, an airport while traveling, or anywhere a modem and personal computer can be used.

This type of model could work well for people who want to set up the equivalent of a 900 number, such as a company that wants to offer consulting services over the Net. The consultants could be qualified people who would alternate with others from the around the world in providing consulting services to customers 24 hours a day.

With a global 800 service, your customers from around the world could call you at no charge. The best part about this idea from the company perspective is that, unlike 800 numbers, for which the receiving party pays the cost, the company pays only for the initial setup and the incremental cost of additional phones.

Internet Telephony and Electronic Commerce

Once we get past the basic demographics of who is using these technologies and why, the next big thing people are waiting for is the emergence of electronic commerce. As commerce emerges on the Internet, spurring significant volumes of purchases, I firmly believe that Internet telephony will be directly responsible for making this happen.

Although there are plenty of people who feel comfortable purchasing products without any direct human interaction, there is an even larger group who demand a "pre-sales close." Internet telephony may very well be the key that unlocks the projected multibillion-dollar electronic commerce industry.

The Growth of Voice Mail Delivered over the Internet

As more people gain access to the Net and start using Internet telephony software products, one of the by-products will be the growth of Internet voice mail. Since the early 1980s voice mail has been popular in corporate America. As Internet telephony software products become more popular, more and more people will use the Internet to leave voice messages for each other. These days, when I get home from work, I check my computer for voice mail messages even before checking my telephone answering machine. There is a chance that in the future voice mail may become even more popular than e-mail.

The Convergence of Interoperablity and Commercial Standards

One of the key items that limits Internet telephony market penetration is the fact that the competing products don't "speak" to each other. And it's not as though standards don't exist. Since the early 1990s, the scientific community has provided leadership and standards for Net-based telephony. However, most commercial companies choose not to follow them. The result is a cyberspace Tower of Babel.

The standards problem is close to being resolved, but until this happens, people who purchase one particular software product cannot communicate with someone who is using a competitor's product. Can you imagine if you purchased a cellular phone from company A and I purchased a cellular phone with different features from company B, and the two of us could not communicate? This is the problem with which Internet telephony consumers are faced. My predication for the near future is that over time most of the major Internet telephony companies will agree on a standard. Agreeing on a standard doesn't mean that a company will give up its own proprietary code. All that it really means is that everybody will agree on the way in which information is shared.

A marketing ploy that seems to be working with very limited results is "buy one, get one free." In other words, if you purchase a copy of WebTalk, for example, included in the box is not only a microphone but a license for a second copy of the product.

The Internet telephony industry is moving toward the adoption of standards. Two major efforts are underway: Netscape/IBM and Intel/Microsoft. Many of the popular Internet telephony software vendors have pledged support for both of these efforts.

Personally, I'm looking to the combined efforts of all the companies involved to unite the Internet telephony industry. With their collective muscle, they can help the industry adopt the standards that allow all companies in this area the chance to gain growth and market share.

Furthermore, as chairman of the Voice on the Net (VON) Coalition, I will do everything that I can with our membership to help bring about the adoption of standards in a timely manner.

The Free World Dial-Up Experiment

It might surprise you that today it is possible for somebody on the Internet to initiate a call to somebody else who is at the other end of a telephone, rather than being on the Internet. Not only is this possible, products such as the Internet Telephony Gateway from VocalTec and Dialogic as well as the Free World Dial-Up (FWD) experiment (see Figure 4.1 and discussion below) have allowed people from around the world to contact friends and relatives using the Internet, even if they don't use a computer.

In addition to my other Internet activities, I helped start the Free World Dial-Up Experiment in the fall of 1995. FWD had its origin in October 1995, on the Internet Phone mailing list. (Internet Phone is a telephony product that is discussed in more detail in Chapter 5.) Of interest to many people on the mailing list at the time was the "patching" of a telephone line into an Internet Phone session. Then Izak Jenie of Jakarta, Indonesia, announced to the mailing list members that not only was a project like this viable, but that he had a working prototype. A few days later, the FWD started. FWD started making this kind of service available to people around the world in February 1996.

In some areas around the world, including the United States, local phone calls are free, or there is a single charge for a phone call of unlimited duration. Fundamentally, FWD provides access to a local calling area directly

FIGURE 4.1 The Free World Dial-Up experiment logo. The home page for FWD is http://www.pulver.com/fwd.

from the Internet. Since 1993, there has been a similar free server available on the Net, called the Internet Faxing Service.

One of the fundamental limitations of Internet telephony to date has been that both parties must be online at the same time in order to establish communication. Basically, this means that two people need to know in advance if they want to meet at a predetermined time to get together for a chat.

FWD was the first software-only solution that eliminated the need for the recipient of the phone call to be online. In the evolution of Internet telephony, this was a significant step forward.

There was a lot of excitement surrounding FWD when it was announced. I received literally hundreds of e-mails from people who wanted to participate in the experiment—mostly from people who had friends and family members overseas and were hoping to find out if there was a commitment for a local server in the town of their family member or friend.

The international media also found the project fascinating, and there was a lot of coverage, especially within the United Kingdom. For instance, in November 1995, on the back page of the *Sunday Times of London,* British Telecom announced that it was investigating the possibilities and implications of the FWD experiment.

During the first couple of days of the FWD project, there were many e-mails and follow-up discussions among the three "founders"—Izak in Indonesia, Brandon Lucas in Tokyo, and myself in New York. We started using PowWow (see discussion in Chapter 13) as our standard for our multiple-user chat sessions. PowWow worked out quite well for our meetings, and it didn't hurt that it's free.

During FWD's start-up phase, we agreed on our mission statement and the answers to frequently asked questions (FAQs) that are still available on the Net today, as well as in Appendix B of this book.

One of the initial challenges was to form a team and delegate the responsibilities accordingly. Within the first week of the experiment, the core project team was assembled. The people who contributed to this effort included Izak Jenie (Jakarta, Indonesia), development; Brandon Lucas (Tokyo, Japan), global server coordinator; Alex Balfour (London, England), media relations; and Lynda Meyer (NYC, USA), legal coordinator. In 1996, Sandy Combs (Burlington, VT, USA) joined the team to assist with media relations. I became the project manager for FWD, set up a few related mailing lists, and helped set up the web pages for FWD as well as the real-time web page for the Global Server Network.

Above all else, FWD was really a social experiment that attempted to determine if our mission could be accomplished by a group of people who had never met and whose sole motivation was to make the Net a better place to be. The FWD project followed the spirit of ham radio in that we allowed only noncommercial users to use the software, which we developed to contact friends and family members overseas. As a social experiment, I believe we were able to prove that people who have never met, from places all around the world, can get together as a team and work toward a common goal.

I acknowledged from the start that FWD was not a scalable technology. What I mean by this is that there needed to be a dedicated PC for each phone line that was going to be made available. If more than one person wanted to make a call into Guam at the same time, then at least two PCs with two separate phone lines needed to be available and online.

When the project started, the project team agreed on a timetable. We put a floating end date of April 1996 (which was extended to June 30, 1996) so that the people agreeing to run FWD servers didn't need to make a long-term commitment to the project. As things turned out, it appears that FWD servers will continue to run past the original project end date and for some time to come.

While the word started to spread, the FWD mailing list became an active discussion group among various participants as well as a collective repository of information. An FWD community formed along the way. Together we learned that in some countries there is no such thing as a free (or unlimited) local phone call; this is the reason there are no FWD servers in certain countries.

When the beta client of FWD was released, the only one experimental server was running in my home. Because I didn't expect many people to have friends in my home town, I published a phone number so that when somebody connected to the FWD server with a phone line, the answering machine came on. During the course of each week, I received phone messages from Australia to Brazil to Pakistan and many points in between from my FWD server. Many of the messages reflected tremendous excitement regarding the FWD project.

The biggest challenge during the first few weeks of the project was finding a commercially available modem that was compatible with the software developed in Jakarta. Not only did I have to find a voice modem that used the Cirrus Logic chipset, but it also (as I learned later) had to have a separate mike input/speaker output. This modem is not used to establish Internet access, but rather to dial a phone number. That is why the speed of the modem—even 2,400 baud—doesn't matter.

Besides Izak, I was the first person to receive a copy of the FWD server software. As soon as I heard that the software was going to be made available to me, I ran to my local computer stores and spent hours looking around for a modem that would work with Izak's software. Little did I know that the true magic of making the FWD server software work hinged on a pair of cables that connected from the voice/fax modem to my sound card.

As a result, the total hardware investment needed in order to set up the FWD server was a $52 modem from Yokohama and a pair of $3.49 attenuator cables.

After my New York server went online, it was followed by a server in Seattle. Shortly after the Seattle server went online, we started seeing servers from all over the world come online as well. During the first quarter of 1996, servers from Australia, Indonesia, Singapore, Moscow, Guam, Sweden, Canada, and the United States went online.

We ended up creating a virtual community of people whose interest was to help extend the virtual community. By and large, the project was a success.

What about Commercial Versions of FWD?

In early 1996, VocalTec, Inc., and Dialogic Corp. announced the development of a gateway product that allows one to interconnect from the public telephone network to the Internet. People have compared the service to something similar to FWD, but this product is a commercial application that has the capability of being scalable.

Corporate use of such gateway products should help speed the introduction and usage of Internet telephony products within the corporate environment. It will be up to the creative and innovative value-added resellers (VARs) to build applications and sell these gateway products as part of their overall solution for a particular company's needs.

If a well-financed start-up company wanted to, it could purchase 100 of these gateway servers and position them in strategic markets around the world. Within a rather short period of time, local regulations notwithstanding, it would be possible to offer a telephone-to-Internet-to-telephone service within these selected major cities. In the future, national Internet service providers such as CompuServe could set up local gateway access servers. Depending upon what action the U.S. Congress or the Federal Communications Commission (FCC) takes in regard to Internet telephony and the interconnection of similar gateway products, you can also expect some large corporations to set up similar gateways as a means to bypass long distance carriers and start seeing immediate cost savings for their long distance phone calls.

What about AT&T and Other Traditional Long Distance Providers?

Recently, AT&T announced plans for its own Internet telephony products and services. In the future, look for all of the major telephone companies to embrace rather than reject these technologies.

When the American Carriers Telecommunications Association (ACTA) filed its petition with the FCC in March 1996 to ban the sale and use of Internet telephony products, my first impression was that ACTA was trying to eliminate outside competition by banning emerging technologies.

What I found really ironic about ACTA's petition was that the ISP with which ACTA has its e-mail account is a supporter of Internet telephony. ACTA and its member firms are acting as though they are "technology challenged." My predication is that the telcos (at least those in the States) will form new media and emerging technology groups that will generate revenue from the Internet.

Will there ever come day when you can pick up a conventional telephone to dial a number overseas and have that call routed over the Internet by your long distance phone company? I'm sure that this is not only possible, but that it will become a reality. If people are willing to have a lower quality of service (with regard to sound and circuit reliability), I can easily envision them choosing to use an Internet long distance carrier rather than what is known today simply as a long distance carrier.

Who Uses Internet Telephony Today?

As discussed in Chapter 1, the size of the Net population that has embraced this exciting new technology has grown rapidly, with members from both the hobbyist, individual, and business communities. During 1995 the majority of users were hobbyists, and one large consulting firm, International Data Corporation (IDC), estimated that 20,000 people a week used this technology. As each month passes, the mixture of hobbyists to individual users to business users fluctuates. Each day there are people who introduce this exciting technology to their friends and family members as well as their business partners.

In addition to using the Internet, some universities are looking to these leading-edge products as something they should use within their own networks or intranets. For example, Cornell University recently announced plans to replace its telephone system with similar technology that allows its computer workstations to act like telephones. The university eliminated the Cornell private branch exchange (PBX) system with a system of integrated computer telephony workstations that are connected to the 16,000 campus locations that used to use a telephone.

Today, if you're a college student, there is a very good chance that you have access to the university T1 line from your dorm room. More and more college students can connect to the Net and use Internet telephony products to speak with family or friends across campus or on college campuses around the world. The technology will probably never replace the telephone, but it may change how the conventional telephone looks. The advent of Internet telephony may result in the global reduction of the costs associated with long distance phone calls. In the future we may see variations of Internet telephony integrated into products like computers, and the dedicated desktop telephone may become a relic of the past.

Leonard Czajka, a friend from Seattle who I met using Internet telephony, once described the calls he made as "the calls you would not have made otherwise." This may still be true among the legions of people who use Internet telephony as a hobby and enjoy speaking with others just for the sheer enjoyment of the conversations. If you have never used a CB or ham radio and can't figure out why strangers would spend time going out of their way to speak with strangers, this type of behavior may seem a bit out of the ordinary. Then again, there isn't very much on the Net that one could consider ordinary anyway. There are many things going on over the Internet that one could find a bit difficult to explain. One of the weirder parts about this communications explosion is that at times I feel as though I know more people from around the world than in my own neighborhood. In many ways, the Net has become a social community unto itself. As friends from my high school and relatives from around the world discover the Internet, and specifically the appeal of Internet telephony, I've stayed in touch with people who I haven't seen (or called) in years.

Years ago I ran a personal ad in *New York Magazine,* which is how I met my wife. Just thinking about where this technology has taken us in the past 10 years, I don't think it would be very far-fetched to expect complete dating services on the Internet in the very near future. Using the right kind of tools, not only would you be able to see "on demand" a preview of your potential "date," but you might be able to click on an icon from a web page and create a real-time virtual link to someone sitting at the other side of that web page. People don't have to give out their telephone numbers to speak with each other, so there's an added layer of safety.

The point here is that as these enabling technologies move forward, they can and will help redefine the way people look at personal and professional relationships. Hopefully, this change in outlook will allow us to displace older technologies with new and better ones.

Chapter 5

Internet Phone: The Market Leader

From February through September 1995, Internet Phone (IPhone) from Vocal-Tec became the *de facto* market leader in an industry that was still in its infancy. To date, three major versions of Internet Phone are available:

- Internet Phone 2.x

- Internet Phone 3.x

- Internet Phone 4.0

As each new version of the product became available, new and exciting features were added. Today, Internet Phone provides features such as full-duplex communications, enhanced audio quality, multiple codec supports (including Truespeech), voice mail, muting, whiteboards, and even a Netscape plug-in.

Internet Phone 4.0 represents a significant change to both the interface and fundamental functionality of Internet Phone (see Figure 5.1). Some of the features available are beyond the scope of this book, but in time—as you become a more experienced Internet Phone user—you will benefit from taking advantage of all the advanced features.

New Features Introduced with Internet Phone 4.0

In addition to the change to the graphical-user interface, VocalTec introduced new features such as the following:

- New dialing and ringing sounds
- More general topics
- A global online directory

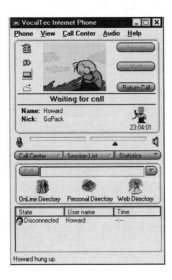

FIGURE 5.1 Internet Phone 4.0.

- A personal directory
- Microphone and speaker muting
- A web directory voice mail
- File transfer
- Text chatting
- A VocalTec whiteboard
- Audio configuration
- Full-duplex communications
- Directory services

Let's examine a few of the more significant changes in greater detail.

IPhone 4.0 Sounds Different

When someone calls you on the telephone, you hear the sound of a ringing U.S. telephone bell. With Internet Phone 2.x and 3.x, the telephone rings sounded like they were recorded from the Israeli telephone system, which, as it turns out, they were. The same hold true for the dialing-out sounds. With IPhone 4.0, these sounds have been styled for the U.S. market.

More General Topics

Instead of one general topic, there are now several. When Internet Phone was introduced, there was just one general topic, which was the "topic" everybody was assigned when first logged into the server. During 1995 the VocalTec engineers realized that if too many people logged into the server and used the general topic, the actual server software could not support all of the people who logged on during a busy period. Busy periods for U.S. Internet Phone users usually occur on Friday and Saturday nights.

At the end of 1995, VocalTec decided to split the general topic into several general topics: GENERAL01 through GENERAL06. Internet Phone 4.0 now calls these areas chat rooms. Internet Phone 4.0 will only allow you to log into one general chat room at a time. So if you are looking around Internet Phone for one of your friends who you know drops by the general chat room, it could take time for you to scan through all the users listed in all the active general chat rooms. Internet Phone 4.0 allows you to connect to only one of the general sessions at a time. If you are looking to meet a friend, the easiest way to find that person is to meet them in a private topic. Otherwise, you need to go through the various general groups in order to find your friend.

The Global Online Directory

The global online directory is your navigation tool for joining any of the public or private chat rooms (see Figure 5.2). To join a particular chat room, all you need to do is double-click on the name of the chat room. As soon as you join, the name of the chat room turns red on your screen, and the following information appears in the bottom of the screen:

User Nickname

Full name

Comment

Origin

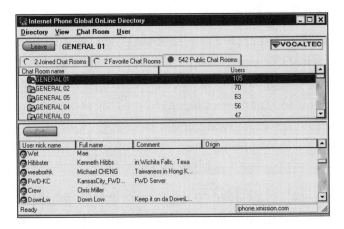

FIGURE 5.2 Internet Phone's global online directory.

Internet Phone 4.0 looks at the IP address of each user and attempts to figure out the host name. For those hosts it can identify, it matches the code and figures out the country of origin. The global online directory is updated automatically by Internet Phone. It does this by communicating with the Internet Phone server you're connected to. The global online directory usually remains open during the time you use Internet Phone.

Muting Capabilities for Microphone and Speakers

A couple of other features from which long-time IPhone users benefit are speaker muting and audio muting. Speaker muting is just like pressing the mute button on a speakerphone—the sound is turned off when the button is pressed. Audio muting prevents the microphone from working when speaker muting is activated.

Voice Mail

Figure 5.3 shows a new feature with the release of IPhone 4.0: voice mail. Internet Phone allows you to record and e-mail a message to anyone. If the person you're leaving a message for does not use Internet Phone, they can still play back your message with VocalTec's voice-mail player, available from VocalTec's website.

FIGURE 5.3 The Internet Phone voice-mail screen.

The Internet Phone Chat Window and Whiteboard

Figure 5.4 shows another new feature, the Internet Phone chat window. You can use this window to type messages to and receive messages from other Internet Phone users. Figure 5.5 shows the whiteboard that comes with Internet Phone 4.0.

FIGURE 5.4 The Internet Phone chat window screen.

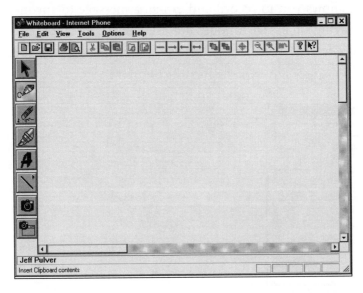

FIGURE 5.5 The Internet Phone whiteboard screen.

Audio Configuration

Figure 5.6 shows the Internet Phone 4.0 audio configuration. With this feature, you can check your audio configuration whenever you change microphones or speakers.

Full-Duplex Communications

As discussed in Chapter 3, full-duplex mode allows you to speak with somebody while that person is talking to you, much like a "real" conversation. This feature really became a marketing tool more than a function that many people would take advantage of. It's true that when you are involved in half-duplex conversations you hear only one party at a time, much like a speakerphone. But unlike a speakerphone, Internet Telephony has some level of delays associated with it—and these delays are anywhere from a half a second to a second and sometimes even more. Because there is a lag time between the time one party starts speaking and the time he is first heard by the other party, how effective do you think full duplex can be? The answer is, not very.

However, full duplex as a feature is useful when there are few delays, and when the Internet has matured a little, some of the built-in delays will begin

FIGURE 5.6 The Internet Phone audio configuration.

to disappear. Then the full-duplex feature will eventually make Internet telephony calls feel more natural.

—— **Note.** *During a full-duplex conversation, it is possible that you could run into a feedback problem. To prevent feedback, I have found that either a headset or handset works well. Products such as those from Jabra and SoundXchange work very well with full-duplex sound cards. The Radio Shack headset model PRO-50 also works quite well.*

Directory Services-Based Approach

Of the major Internet telephony companies, Internet Phone was the first product to use a server for the specific purpose of providing directory services. Many of the other Internet telephony products feature the ability for one party to address another directly.

The server-based approach of the directory services function does provide a live indication as to who is available to speak with you. This is true for most of the Internet telephony providers found in the marketplace today.

Tables 5.1 and 5.2 highlight the differences between Internet Phone versions 3.x and 4.0.

TABLE 5.1 The Internet Phone 3.x Menu Structure

<u>Phone</u>	<u>Options</u>	<u>Help</u>
Call	Auto Accept Calls	
Disconnect	Manual Activation	
Connect to Server	Full Duplex	
Disconnect from Server	User Info	
Register	Hot keys	
Resend Registration Information	Set Activation Level	
Exit	Test System Configuration	
	View as Toolbox	

TABLE 5.1 Continued

Phone	Options	Help
	View Info	
	View Log	
	Clear Log	
	View Statistics	
	Reset Statistics	

TABLE 5.2 The Internet Phone 4.0 Menu Structure

Phone	View	Call Center	Audio	Help
Answer	Session List	OnLine Directory	Voice Activation	Help Topics
Hold	Call Center	Web Directory	Full Duplex	Web Guide
Hangup	Statistics	Personal Directory	SpeakerPhone Mode	Registration Wizard
Do Not Disturb	Status Bar	Outgoing History	Mute Microphone	Support Wizard
Voice Mail	Options	Incoming History	Mute Speaker	About
Whiteboard			Audio Test	
Text Chat			Audio Mixer	
File Transfer				
Exit				

Installing Internet Phone

Before we discuss how to use some of the new features in Internet Phone, let's install it. To install Internet Phone 4.0 for Windows 95, choose the file IPHONE4.ZIP from VocalTec's website, http://www.vocaltec.com. Then follow these instructions:

1. Put your copy of IPHONE4.ZIP in a temporary directory.

2. Run PKUNZIP on IPHONE4.ZIP. All of the necessary files will unpack.

3. Run SETUP.EXE from the temporary directory.

4. The first screen that appears after running SETUP is the Software License Agreement screen. Assuming that you agree with the terms, press Yes.

5. At the Welcome screen, press Next to continue.

6. From the Choose Destination Location screen, you can select the directory into which Internet Phone will be installed. The default location C:\PROGRAMFILES\VOCALTEC\ should be fine for most systems. Press Next.

7. At the Select Program Folder screen, use the default choice, Internet Phone 4.0, as the Program Folder. Press Next.

8. At the User Information screen, enter your personal information:

   ■ Name

   ■ Nickname

   ■ Phone number (be careful here; other users will be able to access this information, so you can leave this field blank if you prefer)

   ■ E-mail address

   ■ Comment

9. Press Next to continue.

10. On the System Information Screen, you need to specify the name of your SMTP server. If you are using either Eudora or Pegasus for your e-mail, you can find out the name of your SMTP server from your e-mail configuration file. Otherwise, you should contact your Internet service provider or system administrator to find out the name of your SMTP gateway. If you leave this field blank, you will not be able to send Internet Phone voice mail.

11. On the System Information Screen, specify your modem speed and indicate whether or not you have a fixed IP address.

Internet Phone 4.0 is now installed on your system. After a minute or two you will be asked, "Do you want the Internet Phone Software to run from Startup?" This is a personal decision, but for my purposes I choose No. The set-up program continues, updating the Windows 95 registry and other system files.

The final screen, Setup Complete, will give you the opportunity to read the latest Internet Phone ReadMe file and allow you to register your copy of the software if you have a licensing code provided from VocalTec. If your copy is not registered, your copy of Internet Phone will work for eight days. For the latest news about Internet Phone, you can drop by the VocalTec website at http://www.vocaltec.com.

Internet Phone Features

To really get up to speed on Internet Phone, you need to take advantage of the following features:

- *User Info.* Be certain to enter as much accurate information as you want others to know about you. Generally speaking, most people put their real names, but never their phone number. This information is readable by the person who you're speaking with over the Internet so be sure that you only include information that you want others to know. In Internet Phone 4.0, this feature is found under Internet Phone Options User.

- *Test Volume Level.* This lets you test the speaker settings. In Internet Phone 4.0, this feature is found under Audio Test.

- *Test Voice Activation Level.* This lets you adjust the voice activation (VOX) levels. Depending on your preferences, you may feel more comfortable using manual activation. Manual activation acts like the push-to-talk (PTT) switch found on CB and ham radio microphones.

In Internet Phone 4.0, manual activation is found in the Audio Voice Activation menu. You have a choice of either Automatic for a Quiet Environment or Automatic for a Noisy Environment.

- *Auto Accept Calls.* Don't enable this feature unless you plan on saying hello to anyone and everyone who tries to contact you.

- *Setting up the Internet Phone audio (for IPhone 2.x and 3.x).* The key here is to set the voice activation to the correct level. This really isn't tricky, but here are the steps you can take to assure success:

    1. Choose Set Voice Activation Level from the Options menu. Once the voice activation level is set, the message "SET VOX" will appear. The level bar will display the recording levels from your microphone.

       Speak into your microphone. The colored bar should respond. This bar indicates the sound level. If it does not respond, make sure your microphone is plugged in and that no other open applications are using your audio board.

    - When you are running in VOX mode, the message "Speaking" will appear on the Internet Phone screen.

Note. *Try not to speak with your mouth very close to your microphone. If the voice level reaches into the red zone, something is wrong with your setup.*

Note. *Internet Phone saves the current voice activation level to disk, so there is no need to repeat this process again unless you change your audio input device or unless it seems the voice activation interface is not working smoothly.*

Setting Up the Internet Phone Audio (for Internet Phone 4.0 Only)

Again, you want to set the voice activation to the correct level.

1. From the menu, select Audio, Audio Test.

2. When the dialog box appears, select Start Test. By pressing Start Test you are able to test your system's audio recording and play-back ability.

Private Topics

Private topics are similar to a private club where you can meet only others who know the club's location. Private topics are handy for business users as well as people who use private calling circles. No password is needed to gain entry to a private topic, but only those people who know about it can find it.

Usually, when you decide to use Internet Phone for private conversations, it's generally because you don't want others to know that you're online. Once you connect to the private topic of your choice, you leave the general session behind.

Connecting to an Internet Phone Server

Before you can actually start using Internet Phone, you need to connect to the Internet Phoneserver network, which is your global phonebook; it presents you with an updated list of topics and online Internet Phone users. You must connect to an Internet Phone server to use the software talk over the Internet.

Connecting via Internet Phone 2.x and 3.x

To connect to any Internet Phone server using IPhone versions 2.x or 3.x, you must first choose Connect to Server from the Phone menu. From the list of servers in the dialog box, select the sever that is closest to your geographical area.

Connecting via Internet Phone 4.0

When you click on Online Directory in IPhone 4.0, the Internet Phone Global Online Directory window opens. Internet Phone automatically tries to connect you to one of the available Internet Phone servers. To select the servers that Internet Phone will choose from, select View Options, then click on Global Online Directory.

Making a Call

Calling someone is very easy. All you really need to do is locate someone from the topics that you've joined and click on his or her name. As you scroll past the list of names, you may see references to their hobbies and related interests, which will help guide you in choosing someone to speak with.

The Call dialog box is used to call specific people over the Internet. It is also used to join topics and see who is listed under them. The first time you run Internet Phone, it lists you under the topic General. With Internet Phone 3.x, you can join more than one general topic at a time. Internet Phone 4.0 will let you join only one general topic at a time.

Calling a Person Using Internet Phone 2.x or 3.x

In order to call someone using IPhone 2.x or 3.x, you need to either click on a Quick-Dial button or choose a name from the Call Dialog Box user list, then choose OK. The image on one of the Quick-Dial buttons will animate and you should hear a ringing sound. Table 5.3 shows some of the Quick-Dial buttons and their functions.

You can receive calls by selecting the button that is animated. Over time you'll notice that people's habits remain consistent, whether they are using Internet telephony or a regular telephone. There will be times when you get a busy signal, and other times when no one chooses to answer your call.

TABLE 5.3 Button Modes for Internet Phone Versions 2.x and 3.x

Button Mode	Meaning in Internet Phone 2.x and 3.x
Not in use.	The button is not in use.
Making a Call.	You are trying to call someone.
Receiving a Call.	Someone is calling you.

Internet Phone User Notes and Registration Tips

When you register your copy of Internet Phone, you need to fill out all of the fields on the form. Otherwise, the Register button won't work. By pressing the Register button, the Internet Phone attempts to connect you to the Vocal-Tec user database.

If you fail to register properly, Internet Phone displays a message indicating that you are not registered. With Internet Phone 2.x and 3.x this occurs after you talk for 60 seconds on a single conversation. Internet Phone 4.0 allows you to evaluate the product for eight days. During the eight-day evaluation period there is no time limit set for conversations.

Make Your IPhone Registration Known

When you register Internet Phone, it creates a file called VCREG.DAT. Whenever you start Internet Phone, the software checks for this file. This is how IPhone knows if the copy you are using is registered.

So, after you register your copy of Internet Phone, you should copy this file to a floppy disk as soon as possible and store the disk in a secure location. Should you decide to change operating systems—from Windows 3.11 to Windows 95, for example—you will need to put a copy of this file in the new Windows directory.

If this file is missing, Internet Phone assumes you are using the unregistered version of the product.

If you bought Internet Phone in a store, the registration code should be on the inside cover of the manual. The number on the blue card is a serial number and cannot be used to register Internet Phone.

Note that you must enter the registration code exactly as it appears in the manual. If you received your registration code by e-mail, you must choose to register From File in the registration dialog box. Make sure not to alter the e-mail message in any way. If you have a problem, be sure to send the information from the file as text, with no attachments, to support@vocaltec.com with a description of the problem. The information in the e-mail registration is for use by the program. The information seen by other users is configurable by selecting User Info from the Options menu in Internet Phone.

Chapter 6

Internet Phone Usage Tips: Questions and Answers

Now that you have an overview of Internet Phone, you might like to have access to some of the helpful information that has been contributed by members of the Internet Phone mailing list. The digests from both the Internet Phone mailing list and the VON mailing list, from inception through mid-1996, are included on the enclosed CD-ROM. You can search these archives for help and answers to questions you have as you use Internet Phone.

This chapter includes answers to some of the more frequently asked questions regarding Internet Phone. Again, if you have any specific questions, please feel free to send e-mail to jw-questions@pulver.com, and I will do my best to answer your questions.

General Internet Phone FAQs

Q. O.K. I'm hooked. I want to order my copy of Internet Phone from Vocal-Tec. I've gone to the company's website and filled out the form. How long does it take to register a copy of IPhone?

A. The actual time varies, usually between 24 and 48 hours.

Q. What is the most current version of IPhone?

A. As of July 1996, it was Internet Phone 4.0. You can drop by the VocalTec home page to find up-to-date information regarding the latest release of Internet Phone.

Q. A friend of mine gave me an evaluation copy of Internet Phone. I just installed it. What steps should I take before I attempt to go "on the air"?

A. If you are using Internet Phone 2.x or 3.x, adjust your voice activation slide knob in IPhone one bar from the left. If you don't, background noise will cause the system to remain on "Speaking." Adjust the voice activation a bar at a time to the right. If you find the activation button never switches back to the person you are talking to due to background noise at your end of the connection, adjust it a bar at a time to the right. Also, leave the Remote Information and Network Statistics window open on your desktop to keep track of who you are talking to and what is going on with regard to the quality of the Internet connection.

When you are looking at the Network statistics, note that the round-trip delay is measured in milliseconds (ms). The average round-trip delay translates into approximately 1,000ms or 1 second. Depending upon the quality of the connection between yourself and the person you are speaking to, the turnaround time might be as low as 400ms or as high as 1,600ms. If you notice that while you speak, your voice level is all the way in the red, adjust your microphone level down or move back from the microphone. Ideally, the voice level indicator should be in the middle of the green area. The proper adjustment of your microphone is critical to your first-time success of being heard when you are "on the air."

Tracing the Route of Your Call

Did you know that it is possible for your voice data packets to travel around the country before arriving at a destination that could be as little as five miles away? A program called Traceroute allows you to trace the route of the data packets. Recently, while a friend of mine from my area and I met on Internet Phone for a 20-minute talk, we both ran Traceroute and noticed that our packets were leaving the New York City area, traveling to Washington, D.C., then across to San Francisco, and then back to the East Coast and to our respective Internet service providers. We both wondered why there was a one-second delay at a time when we felt there shouldn't be any. Little did we know that the Net routes traffics based on the relative address of the server, rather than based on the location of the two parties.

If you are using Internet Phone 4.0, be sure to use the audio wizard to adjust your voice activation levels.

Q. Is it possible to customize the sounds IPhone uses?

A. Yes. This was a very popular thing to do among Internet Phone 2.x and 3.x users. It was possible to change the Internet Phone noises (i.e., ringing, busy, etc.) from the Control Panel>Sound.

Q. The Internet Phone connect and disconnect sound are too loud relative to the sound level I need to use for voice connects. How can I lower them?

A. You can run the relevant sound (.WAV) file through a recorder application to reduce the sound level.

Q. What is the Private IPhone IRC Server Network?

A. This network is a private set of IPhone IRC Servers dedicated to IPhone users.

The current IRCSRVRS.INI file can be found on VocalTec's website. The current list of servers in this network includes the following:

IPHONE.PULVER.COM

IPHONE.VOCALTEC.COM

IPHONE.WAU.NL

IPHONE.INTERSERV.COM

IPHONE.IACCESS.COM

IPHONE.SMARTNET.NET

Internet Phone Bandwidth Notes

Q. How much bandwidth does Internet Phone use?

A. Audio transmitting requires a minimum free bandwidth, because the audio packets can't be slowed down. The original Internet Phone compressed the audio to 7.7Kbit/sec, which is much less than most Internet applications. *Internet Phone never resends lost data.* The Internet Phone uses advanced signal processing algorithms to reconstruct lost packets without retransmitting them, so its bandwidth consumption is constant. The Internet Phone software sends out voice data packets only when somebody is speaking, so it uses the minimum possible bandwidth. However, Internet Phone 4.0 uses the proprietary VocalTec Codec. Each session of Internet Phone can take up to as much as 24K per session. The improved sound quality of Internet Phone came at a cost: the need to increase the overhead in which an Internet Phone conversation takes place.

Q. I've heard from various sources that Internet Phone is the number-one consumer of bandwidth on the Internet today. Is this true?

A. Not at all. In general, the one application that has directly contributed to increased Internet bandwidth consumption is not the Internet Phone, but in fact World Wide Web browsers. During the past 12 months alone, the amount of graphical information downloaded by people using these browsers has increased geometrically. Next in line for bandwidth consumption are FTP and telnet, with e-mail traffic also included on the top-ten list.

The amount of traffic Internet Phone has contributed to the global Internet to date has been negligible.

Q. How does Internet Phone use the IRC servers?

A. When you start Internet Phone, it establishes a connection to an IRC server to announce to the IRC that there is a new user available to converse with someone. The IRC server records the IP address of the user and creates and subscribes to whichever channels the user subscribes to. Once connected to the IRC server, there is some ongoing interaction between the IRC server and IPhone. The loading of the IRC server increases for a moment whenever IPhone users log on and log off.

Q. What happens when two Internet Phone users connect?

A. When an Internet Phone user chooses the name of another Internet Phone user to converse with (and the other party accepts the call), a point-to-point connection is established between the two callers, without involving the IRC server. In the event of a poor connection, unlike most other popular Internet applications, IPhone does not spend any resources trying to resend its own packets.

Internet Phone Network Statistics

Q. Could you please explain in detail the exact meaning of network statistics?

A. The statistics refer only to UDP packets (port 22555) sent by Internet Phone to another Internet Phone and are not related in any way to IRC.

Incoming Packets

Incoming packets are a counter of packets received from the other side of the connection. They can be command packets, "I am alive" packets, or audio packets. It can be very helpful to watch the command and "I am alive" packets. When you call someone and there is no answer, watch the incoming packets. If they move in increments, then it's ringing at the other end of the line, and the remote Internet Phone sends "please wait" packets. If the packets don't move, then the other side can't send you packets (there's no connection, there's a firewall, or the receiver is not running Internet Phone at all). When this happens, you'll know there is no point in waiting for an answer. Note that sometimes it takes several seconds for the first incoming packet to arrive. Wait a few seconds before you assume there's no connection. A busy reply will also increment the counter. When you are speaking to someone (or idle, not listening), the incoming counter should update once every several seconds because an "I am alive" packet arrived. If none arrives within thirty seconds, you will get a "connection timed out" message.

Outgoing Packets

Outgoing packets are the same as incoming, but going from you to the other side of the connection. When you call someone, the counter increments for each call packet. The packets are transmitted endlessly, in several-second intervals, until an answer or a disconnect.

Average Round-Trip Delay

Every ten seconds, the caller (not the call recipient) sends a round-trip packet to the other side of the connection. This is an average round-trip delay. The packet carries the statistics data from the caller to the recipient of the call. The recipient's software then puts its own statistics in the packet and sends it back. In this way, the software on each side of the connection "knows" the other side's lost packets, and the round-trip time can be calculated. Note that only the caller's software calculates the round-trip time, and the recipient gets it as a remote statistic.

Q. How often do the statistics update?

A. The statistics are internally updated all the time, and the display is refreshed every second.

Q. In an Internet Phone call, are we monitoring just the UDP packets?

A. Yes.

Q. Do the "lost packets" represent the lost packets from the person with whom I am speaking?

A. Yes, each lost packet represents the number of audio packets that didn't arrive from the other side. Each packet is assigned a serial number, so we can tell which did not arrive. If your end of the connection loses information because of a slow modem or an old UART, you will most likely receive the message "Send Errors" in the Statistics window, along with the message "Remote lost packets."

Remote lost packets without send errors are errors that were not detected by the software on your end, but somewhere along the path. The problem could lie with the line from the access provider to the backbone, or it could be on any other router along the way, or a slow modem on the recipient side.

Lost packets can be caused by poor performance on the serial port. There is often a great disparity between lost packets on the recipient's end and those lost at the other end. The answer is to have a 16550A UART in your serial port. Windows must also be properly configured to use the FIFO buffers in the 16550.

Q. So, how do I configure Windows to use the 16550?

A. For more information regarding communications port overruns and how to configure Windows to use the 16550, see the 16550.TXT file at my website: ftp.pulver.com/pub/iphone/utils. The16550.TXT file is a document entitled "Understanding Comm Port Overruns," written by Dave Houston. The following are the settings I have in my \WINDOWS\SYSTEM.INI file in the [386Enh] section:

```
COM2Base=02F8
COM2Irq=3
COM2FIFO=1
COM2Buffer=4096
COM2TXSize=16
COM2RXSize=8
```

Of course, the COM port number and base number will depend on what port you're using, as will the IRQ, although these are the standard Windows settings for COM2. The line COM2FIFO=1 is the only critical one for enabling the 16650 UART (it is enabled with =2 if you are a Commt user). It's useful to set your port speed at about double your modem speed (for example, with a 28.8K modem,set 57600 in TCPMAN and 115200 in Control Panel/Ports, COM2). The higher values aren't offered on the ports menu, but you can just enter them manually.

It's also useful to use a better driver than Windows provides. The original COMM.DRV that comes with Windows is inferior for our purposes, although the Windows 3.1 COMM.DRV is a bit better. The best out of the five I've used is CYBERCOM.DRV, available from: ftp://ftp.cica.indiana .edu/pub/pc/win3/misc/. If you use this one, put it in your \WINDOWS directory and change the [boot] section in WIN.INI to

```
comm.drv=cybercom.drv
```

Q. What causes send errors?

A. An error message from the Winsock sendto() function causes send errors. This happens if you are connected via a modem with a bandwidth of less than 10Kbit. Some packets just can't make it because Internet Phone bandwidth (including TCP/IP and SLIP) is 10Kbit. It can also happen with a slow computer (such as a 386).

Q. From time to time I notice that there is a noise on my telephone line. Would having noise on my telephone line effect the remote lost packets?

A. Yes. One of the leading causes of poor performance of IPhone is noise on the phone line. This problem is not an obvious one for many people, because they're used to hearing the noise. Once you realize you have a chronic noise problem and have it repaired, the quality of your Internet connections (and IPhone contacts) will improve dramatically.

Connection Timeouts

Q. I'm having problems with connection timeouts. What does this mean, and what are the possible causes?

A. There are two types of timeout:

- *A "connection timed out" message on the LCD display.* You'll receive this message when you're talking to someone and that person's machine stops responding. The Internet Phone constantly "pings" the other station while you're speaking to make sure that you're still connected. If the other station fails to return "I'm alive" packets for a given period, the connection is assumed to be lost, and you'll receive the "connection timed out" message. This message can also occur if the other side disconnected the call abruptly, possibly by shutting down the computer or hanging up the line.

- *"Server not responding."* This is a dialog message that indicates the IRC server is not responding. Again, Internet Phone's software tries to make sure the connection to the IRC server is "alive." If your dial-up connection is disconnected for some reason, or the IRC server goes down, you'll get the "Server not responding" message. These timeout mechanisms use the Windows timers and are sometimes affected by open full-screen DOS Windows.

Q. Are the "connection timeout" messages related to whether the person I'm talking with is a registered Internet Phone user?

A. "Connection timeout" has nothing to do with the registration. It means that for a period of time (about 30 seconds), no response was received from the other side of the line. The purpose of this message is to make sure you know when you've been disconnected, so you don't continue to talk.

Internet Phone and Sound Cards

Q. I'm having trouble being heard by the other party, even though it appears that all of my settings are correct. What's the problem? I'm using a SoundBlaster card.

A. Look at the Creative Labs Mixer settings for the microphone setting. There are two settings: one is a green light, the other a red one. If only the green one is on, Internet Phone only loops in that mode. The red light shows that the feed is going through, resulting in two-way communication. IPhone users generally rate the SoundBlaster 16 ASP as the best SoundBlaster card available, but without the ASP chip, they rate the SB16 as poor. Number two on IPhone users' list is the SoundBlaster 32Bit AWE, with the ASP chip.

Q. How do I adjust the microphone level of the Media Vision Pro Audio Studio?

A. To adjust the microphone level, open Multimedia Tools and the Pro Mixer and adjust it about two thirds of the way up. Consider adding this Pro Mixer to your startup.

Internet Phone and Microphones

Q. People have been complaining about the sound of my audio. What can you suggest I do to adjust my settings?

A. I would suggest using the recorder software that comes with your Sound-Blaster to record with both your microphone and/or headset microphone so you can compare their qualities.

Q. People are complaining that they hear feedback in my audio. Are there any known fixes for this problem?

A. For microphones that are not unidirectional, open the mixer control. On the bottom right, take the check mark off the microphone output box. That way when you talk you will not hear yourself through your own speakers.

Internet Phone User Comments on Microphones

- "I use the microphone that came with my SB16ASP card, and it seems to work OK. A few beta testers bought the Radio Shack #331060 and it seemed to work OK for a cheap mic, price about $18."

- ". . . A person using Internet Phone must learn how to talk into a mic. It's not the same as a telephone. Your mouth must be at the distance the mic was designed to operate at, and talking at the same volume will also help. My SB mic works best at a half-inch" from my mouth and I don't speak loud . . ."

- ". . . I bought a KOSS M-10 microphone for my pas16. It works great, it has the round wire mesh top with the foam in it. It is a full-sized mic with a nice 3-meter-long cord and has the 3.5mm plug-on it so you need no adapters (it came with a 3.5mm to quarter-inch adapter) . . .I got it for $9.99 at a software store . . ."

- ". . . I've been very pleased with some of the high-end Radio Shack mics. They're made by Shure. The one I use for IPhone cost about $60. Be sure to check your sound card manual to determine the correct impedance. For example, original SoundBlasters (SB1.0 and SB2.0) require 600 ohm (low impedance) mics. Anything other than that just won't work well . . ."

- " . . .I've been very happy with the Radio Shack Pro-50MX head-set/mic combo. They're made by Koss . . .around $60."

Vox Settings

Q. I've noticed that the VOX settings have been troublesome. Is there any way to change the default pause delay from 400ms?

A. Yes. By changing the delay from 400ms to a longer delay like 1000ms, you will be able to speak (and use the VOX) with greater comfort. With the longer delay, you can take a breath while you are talking and not have to worry about having the other person begin talking before you are through.

To change this setting, edit IPHONE.INI (in \Windows), and in the [Audio] section add:

```
MinIdlePeriod=1000
```

to change it to 1 second; if you want .8 seconds, use 800. You can lose words if the VOX is not set correctly or if you start or end your sentences very quietly (or farther from the microphone).

Q. Do most people use VOX or manual activation?

A. The number of people using one method or the other varies from time to time. One advantage to using manual activation is that it gives you a visual sign that the other person has finished speaking (where both are using manual), without having to say "over to you."

—— **Note.** *If you choose to use manual activation, you can use the Enter key rather than your left mouse button to turn manual activation on and off.*

Internet Phone and Speakers

Q. Is the sound quality dependent on the type of speakers I'm using?

A. Yes. People have reported tremendous improvement in sound quality by using Bose MediaMates and other high-fidelity speakers. Others report using the SoundBlaster SB38 speakers; still others report that they get good quality from an old SoundBlaster Pro piped through a boom box and into old stereo speakers.

Internet Phone Utilities

Q. Is there any easy way I can keep track of people I meet using IPhone?

A. You might try using Mitch Hamm's macro recorder utility, which is available from ftp.pulver.com/pub/IPhone/utils/IPhone.rec. This utility, when used with both the Windows Cardfile program and the clipboard icon in the Remote Info window, uses the F12 key to transfer user information from IPhone to the Cardfile. In order to use the IPHONE.REC file, you need to do the following:

1. Download IPHONE.REC from ftp.pulver.com/pub/IPhone/utils in binary mode.

2. Open IPHONE.REC in the Windows (Macro) Recorder. This defines the Ctrl-F12 key.

3. Open the Windows Cardfile program in addition to IPhone.

4. Open the Remote Info window and click on the icon in the lower-right corner.

5. Press Ctrl-F12. The contents from the Remote Info window should now be an entry in the Cardfile.

6. Save the entry.

Q. Are there any Windows programs that display a real-time indication of daytime and nighttime around the world?

A. Yes, there are several. One of the more popular programs is WorldClock, which—when used in conjunction with Tardis—will provide you with a real-time indication of the position of the sun as it moves across the sky.

Firewall Access

Q. Can Internet Phone be configured to operate through a firewall?

A. Internet Phone has no provision for firewall support such as proxy or socks support. However, it is still possible to use it over a firewall by opening the two channels it works with:

- TCP port 6670 (decimal) to remote system (port 6667 is also used by the private IPhone IRC Server Network)

- UDP port 22555 (decimal); all audio is passed through this port on both local and remote machines

Check if these requirements are accepted by your company's security policy. If your site allows outgoing Telnet sessions, it will probably allow Internet Phone usage. This information is valid for the current version of Internet Phone (4.0). Future versions might use different port assignments, but also will have explicit support for firewalls.

Other Internet Access

Q. Does it matter whether I have a SLIP or PPP connection?

A. People have reported that PPP connections result in higher-quality conversations, but SLIP/CSLIP also works.

TIA

Q. Does IPhone work with TIA?

A. Yes, but you need to use TIA 1.04 or greater with the following settings to enable port redirection:

```
tia -r:22555 22555
```

Note. *As an alternative to typing this string, you can make a .TIARC file in your home directory that contains just the line*

```
-r:22555 22555
```

TIA can sometimes be very slow at incoming connections.

Q. Are there any limitations to using IPhone with TIA?

A. Yes. TIA is not a real IP connection. The entire Internet "sees" only the host UNIX system. It's just as though the software you're running on your machine were running on the UNIX system. This implies several limitations to using TIA with Internet Phone:

- *Only one user of TIA on a UNIX host can use Internet Phone.* If another user tries to use it, the result is unpredictable. The "control" of this one connection could even be transferred between them.

- *It is possible that you can call one user but connect to another.* If both users use TIA on the same machine, and just one is talking, you will be connected to that person, even if you didn't call him.

- *The IRC connection works just fine.* The IRC uses TCP protocol, which TIA can handle. The actual voice connection uses UDP, which causes the problem.

CU-SeeMe and Internet Phone

Q. Do CU-SeeMe and IPhone work well together using dial-up modems?

A. Yes. Quite a few IPhone users use CU-SeeMe. Recommended settings for video are as follows:

Speed	Max kbits/sec	Min kbits/sec
14.4	3 or 4	1
28.8	3	1

Internet Phone and OS/2 Warp

Q. Does IPhone work with OS/2 Warp?

A. Yes. You can run IPhone in a WinOS/2 window.

Q. Is there anything special I should do before trying to use IPhone?

A. Yes. Be sure to add the following line to your AUTOEXEC.BAT file:

```
SET ETC=X:\TCPIP\DOS\ETC
```

Replace the X with the drive that TCP/IP is on. Replace the directory with the settings from your machine.

The Internet Phone Mailing List

Q. How do I subscribe to the IPhone mailing list?

A. It's simple. Send an e-mail request to majordomo@pulver.com. Leave the subject blank, and in the body of the message, write "subscribe iphone."

Q. Now that I've subscribed to the mailing list, how do I send mail to the list?

A. Address your mail to iphone@pulver.com.

Q. Is there a digest version of this list available?

A. Yes. To subscribe to the digest version, send an e-mail request to major-domo@pulver.com. Leave the subject blank, and in the body of the message, write "subscribe iphone-digest."

Q. Now I'm on both mailing lists. How do I unsubscribe from one?

A. Send an e-mail request to majordomo@pulver.com. In the body of the message, write either " unsubscribe iphone" or "unsubscribe iphone-digest," whichever applies.

The IPhone User Directory Web Service

Q. How do I submit my URL to be added to the user directory?

A. Send your URL with your IPhone nickname and preferred e-mail address to webmaster@pulver.com.

Q. How long does it take to have my directory entry added to the user directory?

A. The current turnaround time is 7 to 10 days.

Q. My e-mail address has changed. Who should I contact?

A. Please direct all additions, changes, and updates to webmaster@pulver.com.

Chapter 7

WebPhone

WebPhone 2.0 from NetSpeak Corporation (see Figure 7.1) was one of the first products to implement the following innovations:

- A graphical user interface (GUI) with a Windows 95 look

- A design that had the appearance of a cellular phone

- An advertising model that allows you to see ads running on the WebBoard while you are in a conversation with someone

When I first saw WebPhone back in August 1995, it was the second of what had become a field of more than twenty products that offered Internet telephony functionality. I remember having a very positive first impression, if for no other

FIGURE 7.1 The WebPhone.

reason than that the GUI was excellent. It impressed me because it is a product that offers the capabilities of a telephone and even looks like one.

With WebPhone, the only thing that I had to get used to was the Windows 95 GUI running in a Windows for Workgroups development environment. If you have yet to use Windows 95 and don't know that you need to click on the X in the top-right corner of an application in order to exit that application, I can assure you that you are not alone.

When WebPhone 1.0 was announced, NetSpeak indicated that more than 40,000 people had taken part in the beta test. I wasn't surprised, given the quality of the final product. My only real surprise was that it wasn't named as the best-looking Windows application for 1995. Among the major innovative features of WebPhone 1.0 were the following:

- Voice mail

- Four-line capability

- Caller ID

- Do not disturb

- Speed dial
- Call holding/muting
- Conversational encryption

As you try out some of the Internet telephony products, you will begin to see a difference between the approach of some software developers in the implementation of their respective products. Some products really make me feel like I'm visiting a "hang-out spot" where I know many of the people—the regulars. Then there are others that make me feel like I'm a guest at a private club and that I need to know the name of every person I'm visiting.

Because WebPhone doesn't currently approach directory services the same way other products do, it may take you a couple of minutes to scan the users who are currently online and available to talk to you. Once you get used to WebPhone, however, I think you'll be as impressed as I was.

System Requirements for WebPhone

The system requirements NetSpeak suggests for WebPhone are the following:

- A 80486DX-33MHz computer running Windows 3.1 or higher
- 8MB of RAM
- An MCI-compliant sound card
- A Winsock 1.1-compliant stack
- A 14.4KB modem with compression and error correction turned on
- A VGA card capable of displaying 256 or more colors
- A full-duplex audio card (required for full-duplex communications)

Here are my suggestions for the minimum system requirements for WebPhone:

- An 80486-66MHz computer or faster (a Pentium is always nice)

- At least 8MB of RAM if you are using Windows 3.1; 16MB of RAM if you are using Windows 95

- At least a 19.2KB or faster network communications link

- A WinSock 1.1 (or higher) compliant sockets library

- A Windows-compatible sound card that supports 8KHz or 11KHz sampling (the SoundBlaster 16 works well; if you need an 8-bit card, try the SoundBlaster 2.0)

- A VGA display card that supports 256 or more colors

- A handset or a headset if you plan to run full-duplex mode; without one, there is potential for a feedback problem with your system

- 6MB or more of available hard-disk space

WebPhone works with the following operating systems: Windows 3.1, Windows for Workgroups 3.x, Windows 95, Windows NT, and OS/2 Warp with Windows.

Installing WebPhone

To install WebPhone:

1. Run WP201B1.EXE from the Windows Program Manager.

2. Enter the name of the directory in which you want to install the software (C:\WebPhone) and click on Install.

3. When the license agreement appears, read it; if you agree to it, click Accept.

Getting Started with WebPhone

Before you can establish contact with anyone else, you need to enter information in the WebPhone Configure window, as shown in Figure 7.2. You use

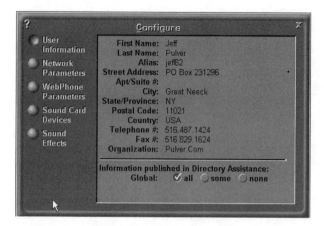

FIGURE 7.2a The WebPhone Configure window.

the Configure window to enter your user name, e-mail address, password, and the address of both your SMTP gateway and point-of-presence (POP) server. You *must* enter the correct information. If you don't, there is a very good chance that WebPhone will not behave properly. The SMTP gateway needs to be specified in order to be able to send WebPhone voice mail. The POP server information is needed to receive Webphone voice mail. In addition, if you don't enter accurate information, it is possible that your entry in

FIGURE 7.2b The WebPhone Configure window.

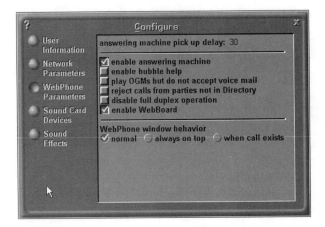

FIGURE 7.2c The WebPhone Configure window.

NetSpeak's Directory Assistance will not be correct which means that other WebPhone users may not be able to find and contact you.

To help you enter the correct information the first time, you should understand what all of the parameters stand for:

- *E-mail:* Used by WebPhone as your "phone number"
- *WebPhone Password:* Used by WebPhone to prevent others making unauthorized changes to your information

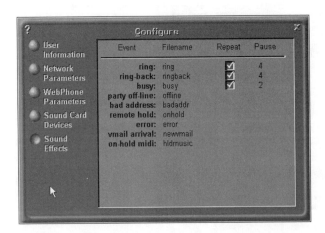

FIGURE 7.2d The WebPhone Configure window.

- *POP server address:* **N**eeded so you can receive voice-mail messages: the POP server is responsible for receiving voice mail from others

- *SMTP server address:* Used for sending voice mail to offline parties; the SMTP gateway is responsible for sending the voice mail to the intended recipient

- *E-mail login:* Used for connecting to your POP server

- *E-mail password:* Used for connecting to your POP server

As far as I know, Netspeak does not do anything with your password information, and many people feel secure about it. If you are really concerned about the integrity of your account, do not disclose your password to anyone. If you choose not to disclose your password in the Configure window, it won't affect the performance of WebPhone, but you will be prompted to enter your password each time you check for voice mail.

In case you are wondering what to enter in the SMTP gateway and POP server parameters, the answers can be found in your configuration file in your e-mail or Internet Mailing software package. If you are unable to figure out the correct values, you should contact the customer support department of your ISP for additional assistance.

Here's an example of a set of WebPhone network parameters. Let's assume that the name of the business is Pulver.Com and that the user has chosen to include his password.

```
E-mail address:       bus-dev@pulver.com
POP server address:   mail.pulver.com
SMTP server address:  mail.pulver.com
E-mail login:         busdev
E-mail password       *******
```

Once you have entered the appropriate user information and associated network parameters, click on Configure. You're ready to start!

Finding Someone to "WebPhone" With

WebPhone differs from most of the other products covered in this book in that WebPhone doesn't give you an immediate and real-time indication as to how many people are online and which topics they are using. However, you can search the directory and find out how many people are online who match your search criteria.

To search the WebPhone directory to find someone who is online, these are the steps you need to take. Let's say that you are looking for someone in New Jersey:

1. Click your mouse on WebPhone and bring it into focus.

2. Click on the Inf button, which is located on the bottom left of the screen.

3. Click on Only Parties Online from the Information screen.

4. Click on the state field of NJ.

5. Click on Search.

After a few moments, the search results are displayed. If there are no matching records, a message indicating this fact will appear.

The WebPhone approach to the marketplace has been to go after the business user. Although the number of business people using Internet telephony in 1996 is a small part of the current user base, I am a firm believer that this market segment will grow—and could grow considerably within the next year. Here's why I believe that: If Internet commerce is really going to take off, there will have to be an Internet telephony component to it. In order to close sales, many people need the "pre-sales close;" others want to speak with somebody in customer service, track the status of an order, and so on. WebPhone is well positioned to make significant inroads to the business community because of its design and operation, which meets these customer needs. The look and feel of WebPhone set it apart from the rest of the products as a business tool rather than just another chat tool. Also, one of the key features—conference calling—makes this product more appealing than the other products for business purposes.

WebPhone was also one of the first Internet telephony products to support direct user-to-user dialing without the need of using an IRC-like chat server. The WebPhone directory is server based, which means it does keep track of all of the users who are logged in at any given time. Although WebPhone uses its own server to keep track of who is online, it does not display the list of users to you on a real-time basis, as many of its competitors do. If you want to know if a specific individual is online, you need to search the directory for a specific match. It seems clear that WebPhone is targeting an audience of people who know in advance that they want to speak with each other, rather than people who casually meet, which is more commonplace with the other products.

WebPhone's directory assistance feature is shown in Figure 7.3. This directory helps you find someone you can talk to. Assuming your WebPhone is configured and it is running on your desktop, the following are the steps you need to take in order to access WebPhone directory assistance:

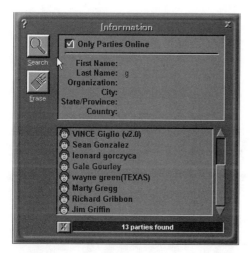

FIGURE 7.3 WebPhone's directory assistance feature.

1. Click on the Inf button. This displays the Information dialog window.

2. Click on Parties Online.

3. Enter your search criteria. For example, enter the letter J in the First Name field and you will find everybody who is online whose first name begins with J.

4. Now press the Search button. The search begins. On the bottom right of your screen, the status of your search is indicated. The maximum number of matches you will see returned from any search is 100.

5. When the list of matches appears, scan it by clicking on items in the list with your right mouse button to see the user details about each person. Double-click on the name of a person with whom you wish to establish contact.

When looking for someone to talk to, you can search on the following items:

- First name
- Last name
- Organization
- City
- State/province
- Country

It Just Keeps Getting Better

The feature sets of all of these emerging Internet telephony products keep changing and improving with time. During the initial beta testing of WebPhone, the developers used the VON mailing list to solicit feedback from potential customers. They also used the list to monitor the reactions of the beta testers. While other software companies also used the VON mailing list, the folks at NetSpeak were the most proactive in responding to specific issues raised by users and converting these issues into new product features in later releases. The result is a better product.

WebPhone 1.0 provided the following features:

- Telephone quality, real-time speech
- Point-to-point calling via e-mail or IP addresses rather than via IRC servers
- Full-duplex operation
- Lines of simultaneous conversations
- Call holding, muting, and blocking
- Last-party redial
- Complete caller ID
- Speed dialing
- Conversation encryption
- Personal phone directory
- Integrated real-time directory assistance
- Integrated voice-mail system for sending and receiving voice mail
- Party-specific, user-definable, custom outgoing messages
- Interactive multimedia user manual
- User-configurable sound effects
- Operation over the Internet as well as any private TCP/IP-based network.

When WebPhone 1.0 was in beta, it helped set user expectations for the second-generation, Windows-based Internet telephony products. As a leading-edge software development company, NetSpeak was very responsive to adding features sought by users to WebPhone 2.0. As an example, WebPhone 2.0 provides the following additional features:

- Call conferencing
- Offline voice mail
- Call transfer
- Party-selective text chat NoteBoard
- Party-specific call blocking
- Party-specific ring through do not disturb (DND)
- On-hold musical instrument digital interface (MIDI)
- WebPhone password for security

Let's take a closer look at some of these features.

Call Conferencing

Your friends may tease you that you've taken a $2,000 computer and, with the right hardware and software, helped make it perform like a $20 telephone, but nevertheless, these WebPhone features are impressive if you're interested in Internet telephony. Call conferencing is one welcome new feature. Just as you can with a telephone conference call, you can use WebPhone to speak with two or more people on one line.

For example, if you need to put together a conference call with you, your sister Risa, and your brother Farrell, let's assume that you have already connected with Risa on line 1 (L1) and Farrell on line 2 (L2). To conference everybody together, move the mouse onto the L1 button and drag it onto the L2 button. Line 1 should go idle, and line 2 will change to C2, to indicate that a conference is currently taking place between the people on L1 and L2. To add another person—your father, Bernie, for instance—you need to connect with Bernie on line 3 (L3), then click on L3 and drag it over to the C2 button. If you want to add more people beyond three, others who are also participating in this conference call (like Risa or Bernie) can include them by following these same steps. To end the conference call, simply press the End button.

Although you may have a full-duplex sound card in your computer, the conference calls on WebPhone are half duplex. Because the calls go through one computer, I strongly suggest that the person with the fastest computer and biggest available Internet bandwidth be the person who initiates the calls.

Offline Voice Mail

If you try to call somebody who is not available because they are offline, it's possible to leave that person a voice-mail message telling him you're trying to reach him. You can actually record a voice message and send the recording to your friend's e-mail address. The message you leave is sent to the person over the Internet. When the person you are trying to reach is back online, they will be able to retrieve the message.

How to Send Voice Mail

When you use WebPhone to contact someone whose answering machine is off or if the party is not available or offline, the WebPhone voice-mail Composer dialog box will appear. In order to record your message, do the following:

1. Press the O (Record) button and record your message.
2. Press the Stop button when you are finished recording.
3. Press the Next button to hear your recorded message.
4. Press the End button or select Exit from the voice-mail composer to send your recorded voice mail.

Note. *If you don't want to send the recorded voice-mail message, press the Cancel button prior to pressing the End button.*

How to Retrieve Voice Mail

You can configure your e-mail programs to automatically grab any WebPhone voice-mail messages automatically. Simply start WebPhone before you begin your e-mail program. Programs such as Eudora and Pegasus can intercept your offline voice mail and treat it as though it were a

MIME attached document. You can also use File Manager and drag the voice-mail attachment from the file system into the Voice Mail Messages window.

Call Transfer

With WebPhone, you can transfer a call to another person as long as that person is in your personal directory. If you want to transfer a call from your-self to someone else, you also need to have an entry for that person in your personal directory. Let's assume that you are speaking with your sister, Lauren, and she wants to speak with your other sister, Michele. Both Lauren and Michele have entries in your personal directory. To transfer your call with Lauren and Michele, you would do the following:

1. Press the Dir button to open your personal directory.

2. Hold down the Alt key and drag the Line button, which contains your call with Lauren, to Michele's entry in your personal directory.

Party-Specific Call Blocking

At home or at work, wouldn't it be nice if you could actually set up filters so your phone doesn't ring if the person who is initiating the call is someone you don't want to speak with? On the Net, this is not only possible, it is one of WebPhone's better features. In fact, with WebPhone you can either route certain calls to your answering machine or reject them altogether so that they go into your voice mailbox.

Although WebPhone allows you to block calls from a specific party, you do need to have that party entered into your personal directory before WebPhone can act upon your block. Let's assume that you want to block calls from Susan Goldman. If Susan isn't already in your personal directory, you need to add an entry for Susan. To do so, follow these steps:

1. Click Dir to go to your personal directory, then, in the "G" section, add Susan Goldman, with all of the information you know about her.

2. At the bottom left of the screen, select Reject Call.

3. Click the Update Party text from the top of the dialog box to save your change.

It is also possible to configure WebPhone so that you receive only calls from people listed in your personal directory. To do so:

1. Click the Cfg button to open the Configuration window.

2. Click the WebPhone parameters button to display the current WebPhone parameters. By selecting the Reject Call button, you will set your WebPhone to accept calls only from people who are already in your directory.

Party-Specific Ring-Through Do Not Disturb (DND)

With WebPhone it is possible to permit some people to ring through, even though you have enabled the "do not disturb" function.

When you enter someone into your personal directory, one of the options is Ring Through Do Not Disturb (DND). In other words, even though you have the DND enabled, it's possible to have WebPhone allow certain people access to you anyway.

To enable the do not disturb function, click on the Dnd button. The DND annunciator appears in the WebPhone display. To disable DND, click on the Dnd button again. The DND annunciator disappears from the WebPhone display.

On-Hold MIDI Music

First, WebPhone provided the ability to put a caller on hold. Now you can even play music for them while they wait.

MIDI, which stands for musical instrument digital interface, has been very popular among electronic musicians since it was developed. Before the advent of RealAudio, MIDI was one of the more popular audio formats found on the Internet, together with .WAV and .AU files.

You can find several Internet reference sites on MIDI. One of my favorites is http://www.eeb.ele.tue.nl/midi/index.html. Another good source for MIDI is http://www.uwm.edu/~jwb/midicol.html.

To specify the MIDI sound, you need to open the Configure window from the Sound Effects section. You can specify any MIDI file as the default on-hold music (HLDMUSIC.MID). Drop by the WebPhone home page for MIDI sound clips, which you can substitute for the default sound clip that came with your copy of WebPhone.

WebPhone Password

The password you specify for WebPhone prevents another WebPhone user from using your e-mail address. When you specify a password, you should use one you will remember. You can change the password any time.

WebPhone Databases

WebPhone uses several files to keep track of user-specific information. It's important to back up these files from time to time so that if you have a hard-disk crash, your personal configuration options and personal calling directory will remain intact.

WebPhone uses three databases while it's running. These databases are the database of received voice mail, the WebPhone directory, and the database of custom outgoing messages. These databases are maintained in the following directories:

- The WebPhone directory database is located in the file WEBPHONE\WEBPHONE.DIR.

- The database of custom Out Going Messages (OGM) is located in the file WEBPHONE\OGM\OGM.DIR.

- The database of received voice mail is located in the file WEBPHONE\VMAIL\VMAIL.DIR.

- The WebPhone configuration information is located in the file WEBPHONE\WEBPHONE.CFG.

To avoid inconvenience in the event of a hard-disk crash, you should back up these files every once in a while (at least once a month). While you're at it, back up the WebPhone configuration file as well.

Chapter 8

WebTalk

WebTalk from Quarterdeck is another product that provides you with the ability to talk over the Internet (see Figure 8.1). Quarterdeck's marketing department immediately recognized the fact that although most computers sold in the United States these days have sound cards and speakers, they're missing a microphone. Therefore, Quarterdeck decided to include a microphone in each box of WebTalk.

But that's not all. In addition to the microphone, WebTalk also contains a second license to give away to a friend, relative, or business partner, so the buyer can have somebody to talk with online. Quarterdeck was the first Internet telephony vendor to bundle a second license in its product package. Many

FIGURE 8.1 The WebTalk screen.

competitors followed suit. In addition, WebTalk—even the evaluation copy, which is available from the Quarterdeck website—comes bundled with Quarterdeck's version of Mosaic, Qmosaic as shown in Figure 8.2.

Quarterdeck Corp. was one of the original licensees of the Mosaic source code. Bundling QMosaic with its telephony product ensures the dissemination of the Quarterdeck name.

FIGURE 8.2 The Quarterdeck browser, QMosiac.

Another component of the WebTalk package is Quarterdeck's Internet Connection software. If you purchase WebTalk in a computer store and you're not on the Internet yet, the Internet Connection provides an easy-to-use menu that allows dial-up Internet connectivity.

Currently, WebTalk does not allow the user to talk with more than one person at a time, but it does provide a text chat window. Before, text chat windows have been added to most of the Internet telephony products covered in this book, although the real benefits of this feature are sometimes hard to grasp. In the future, as additional competitive features are added to WebTalk, it's a safe bet that conference calling and whiteboarding will be supported.

During 1995, Quarterdeck was part of the small list of companies that morphed themselves into an Internet software company. So far, it looks like this shift is working.

Introduction to WebTalk

WebTalk falls into the category of Internet telephony products that take advantage of directory services features. In Chapter 1, we discussed the fact that directory services are dedicated servers with which an Internet telephony product connects to determine who is online or available to talk. Quarterdeck uses its own proprietary set of Internet relay chat (IRC)servers, which the user needs to connect to in order to find others to speak with. With WebTalk, users can also bypass their server network and directly connect to another person.

The design of WebTalk is pretty straightforward. Unlike some of the other telephony products such as FreeTel and WebPhone, WebTalk's microphone controls are directly accessible from the main screen—rather than in a hidden window, as shown in Figure 8.3. Generally speaking, once you adjust your microphone and speaker setting, you will not have to touch them again.

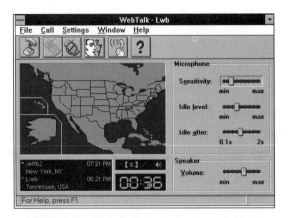

FIGURE 8.3 The WebTalk main screen.

On the WebTalk main screen, you'll see icons that provide you with the following functions:

- Hang up the call
- Display the chat window
- Hide the chat window
- View/change your personal settings
- Mute the microphone
- Access the WebTalk online help

The most noticeable difference between WebTalk and the other products discussed in this book is the World Map feature, which is part of the WebTalk main screen. In addition to helping you speak with people all over the world, WebTalk's World Map feature draws a line on the map between your location and the location of the person you're speaking with, every time you make a connection. Think of it as a free geography lesson.

Getting Started with WebTalk

One of the confusing parts of getting WebTalk installed is Quarterdecks's use of its proprietary Winsock.DLL, which the company calls QWinsock.

QWinsock can be used in addition to the Winsock.DLL that should already be installed on your system. If Winsock.DLL is not currently installed on your system, you can choose to use Quarterdeck's QWinsock as your default Winsock.DLL. In order to run any TCP/IP windows program, your computer needs to have Winsock.DLL installed.

Installation Directions

To install WebTalk, you need to first run the program called INSTALL.EXE. As the installation program proceeds, you will be asked to verify certain parameters. As each installation screen appears, you need to press the Next button to move to the next screen, which shows you accept the parameters outlined on the screen.

The screens you will encounter include the following:

- *Welcome to Quarterdeck WebTalk.* You don't have to enter any information on this screen.

- *Personalize Your Quarterdeck Product.* You need to enter your name, company (if appropriate), city, state, country, and the serial number of your copy of WebTalk.

- *Installation Options.* You need to specify the base directory path to which WebTalk should be installed. The default drive is C:\INET. You need approximately 5MB of available disk space to complete the installation.

- *Installation Components.* You can select the path to which Quarterdeck Mosaic Browser, WebTalk, and Location Manager will be installed.

- *QWinsock Installation.* This screen checks whether or not you have an existing Winsock.DLL in your path. If the installation program finds an existing Winsock.DLL, you are given the option of using either QWinsock plus your existing Winsock.DLL or only QWinsock.

- If for whatever reason, you don't have an existing Winsock.DLL in your path, you should choose to install QWinsock. Otherwise, I would not recommend installing QWinsock.

- *Finishing the Installation.* This screen should display the message, "Quarterdeck WebTalk installation has completed successfully!" If it does, click on the README button so you can look for any recent changes to WebTalk.

You might see a message, "Video for Windows Required," which means that you need to install the Video for Windows driver. If you don't see such a message, you can click on EXIT to leave the WebTalk installation program.

You will need to restart Windows before the changes you made during installation can take effect. You can choose to restart Windows now or select "Don't Restart."

Once WebTalk is installed, you need to log into the private WebTalk Server Network. The WebTalk Server Network is the "place" where you will meet other people from around the world who use WebTalk. Once you are logged in, you can select from the WebTalk White Pages user directory or visit any of the "rooms" in which people meet to discuss topics of mutual interest.

When you enter a room, you are presented with a list of people who are already in the room and who are available to speak with you. In order to talk to people using WebTalk, you need to follow these steps:

- Start your browser and enter the URL for the WebTalk home page: http://webtalk.qdeck.com/.

- Log into Quarterdeck's WebTalk Server. The first time you log in, you are taken through a registration process. Choose a user name and password that you will remember. During the registration process, you are asked to enter the serial number of your copy of WebTalk. The serial number is located on the first diskette enclosed in the package. You can also find it by clicking on the WebTalk icon and then selecting Help About. The serial number for your copy of WebTalk will appear in the dialog box.

- Select someone to speak with.

Like Internet Phone, WebTalk uses a private IRC server. And like Internet Phone, WebTalk uses the IRC server to associate your unique Internet address (your IP address) with a nickname and/or your e-mail address. Of course, if you have a dedicated connection to the Net or have arranged with your Internet service provider to obtain a static SLIP/PPP account, your IP address is predictable.

Both Internet Phone and WebTalk note your currently assigned IP address, together with the information you supplied as your user information. Each time you log into the network, this information becomes available to others who are scanning the real-time directory lists in search of somebody to speak with.

Logging In

After starting your web browser (assuming you have one; otherwise, use QMosiac), go to http://webtalk.qdeck.com.

The first time you log in, you will need to register yourself on Quarterdeck's Virtual Meeting Place User Network. As a first-time user, you will need to supply the following information:

- Your name
- Your WebTalk serial number
- Your WebTalk nickname
- A password

The next time you log in, all you will need to provide is your nickname and password.

When you register your nickname, Quarterdeck actually stores the nickname and checks within its own databases to make sure that the nickname you supplied is unique. In the event you supplied a nickname that someone else is already using, you will be asked to resubmit your proposed WebTalk nickname.

As a way of adding to the community feeling among WebTalk users, you can also supply an e-mail address and the URL to your home page. This way, people you meet on the air can easily follow up with you after your conversation is over, and they will have a chance to "check out" your home page.

Using the WebTalk User Directory

When you log onto the WebTalk user directory, as shown in Figure 8.4, you are presented with a graphical representation of the real-time environment. On the left side of the screen you see a listing of all of the available rooms, and on the right you see the details describing the people who are in the various rooms. As you'll remember from Chapter 1, products with directory services generally allow users to create "topics" or "rooms" that people can "enter" or "join" for the specific purpose of finding others who share the same interests. With WebTalk, a room is called a chat area.

To enter a room, all you need to do is to double-click on the name of that room on the left side of the screen then click the Enter the Highlighted Room toolbar.

The following functions are available by selecting the appropriate icon from the user directory:

- Create new room
- Enter highlighted room

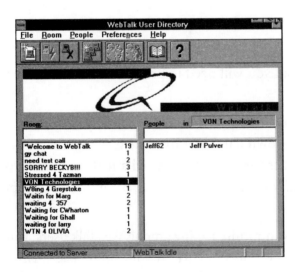

FIGURE 8.4 The WebTalk user directory.

- Exit the room you are in
- Update the list of rooms
- Display user information
- Call the person highlighted on the right side of the screen
- Display your personal user directory
- Access the WebTalk online help

Unlike Internet Phone, WebTalk does not update the current list of rooms and associated users on a real-time basis. If you want to see an up-to-date snapshot of the list of rooms and users, you will need to press the Update icon.

Setting Up a Room

If there aren't any rooms representing things you want to talk about, you can set up your own. Let's suppose you are interested in running and would like to speak with others about that topic. Simply create a new room called "Running" and wait for people to drop in to speak with you.

Identifying Speaking Partners

If you want to find out information about a particular person on the user list, you can highlight the nickname and click on the icon that represents "display user information" or you can simply double-click on the user's name. Figure 8.5 shows a dialog box containing a user's information.

Making a Call

To call someone from the list of people within a specific room, you need to do the following:

1. Highlight the user you want to call.
2. Click on the Call icon on the toolbar.

After you connect with another WebTalk user, the first thing you see is a line drawn on the WebTalk map from your geographic location to the location of the person you are speaking with.

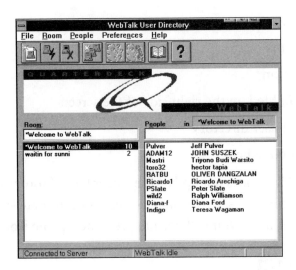

FIGURE 8.5 User information in a WebTalk dialog box.

WebTalk also supports a personal directory in a way similar to WebPhone. You can add people to your personal directory and simply click on their names to try to contact them. However, unlike WebPhone, if the party you are trying to reach is not available, you cannot leave them voice mail.

Bypassing the Server

It is possible to directly contact another user rather than use the WebTalk server network. In order to connect with that other person, you need to know his or her IP address.

The following are the steps you need to follow to establish a direct contact:

1. As you are running WebTalk, click on the Place Call icon from the Call menu.

2. When the dialog box appears, enter the IP Address of the person you want to contact.

3. Assuming the person you are trying to reach accepts your call, you should now be in contact with that person.

Chapter 9

TeleVox

One of the most eagerly awaited Internet telephony products to hit the Internet in 1996 was TeleVox (see Figure 9.1). TeleVox began life as CyberPhone, developed by a University of Pennsylvania student, William G. Foglesong. TeleVox was originally a free multiplatform Internet telephony product. In 1995, when I conducted a survey to determine the most popular telephony product of that year, CyberPhone finished in a very respectable third place. Voxware, Inc., obtained the rights to CyberPhone in late 1995 and relaunched the product as TeleVox.

FIGURE 9.1 The TeleVox main screen.

Voxware is committed to providing TeleVox for free and in multiplatform form. TeleVox has become one of the ways that Voxware has chosen to help showcase its line of very efficient codecs (compression/decompression routines).

One of the features that sets TeleVox apart is that its users can communicate with any LiveMedia-compatible Internet phone. LiveMedia from Netscape Corporation is an open framework for building real-time audio/video applications for the Internet. Netscape used Insoft's technology to create the Netscape LiveMedia framework. When LiveMedia was announced, eleven companies came on board to support it, including Progressive Networks, Adobe Systems, Digital Equipment Corporation, Macromedia, Netspeak, OnLive!, Percept, Silicon Graphics, Inc., VDOnet, VocalTec, and Xing.

The Netscape LiveMedia framework is based on the Internet Real-time Transport Protocol (RTP), RFC number 1889, and other open audio and video standards such as MPEG, H.261, and GSM. These standards enable products from various companies to work together seamlessly and provide users with a range of real-time audio and video capabilities on the Internet.

Netscape has published the LiveMedia framework on the Internet and has openly licensed key technology components of it. Netscape continues to work with the Internet standards bodies to facilitate the adoption of this technology as a formal Internet standard.

Today, various audio-compression routines are found in Internet telephony products. Some of the most popular codecs include Voxware's RT24, DSG Group's TrueSpeech, and GSM. None of the codecs have as a high a compression ratio as Voxware does, at 53:1. By comparison, the compression radio of TrueSpeech ranges from 15:1 to 27:1.

I have found that Voxware's RT24 codec is the most efficient in the marketplace today. This means you can obtain high-quality sound while using just a fraction of the bandwidth other codecs require. For example, Voxware's RT24 requires just 2.4K of bandwidth for audio. In comparison, the original codec developed by VocalTec for Internet Phone required 7.7K of bandwidth just for audio. WebPhone, which uses the TrueSpeech codec, requires 11K of bandwidth.

Voxware's MetaVoice is the Netscape-endorsed technology for low-bandwidth speech transmission with Netscape Navigator. This technology could position Voxware as one of the Internet telephony leaders, depending on how well Netscape is able to penetrate this market niche.

If you are looking to have a little fun, TeleVox also lets you change the way your voice sounds. I'm not sure how practical this might be during phone conversations, but the effects of the product's VoiceFonts are pretty cool. Experiment with a font called Gender, or try Robot. Make it sound like you're talking from a cave or a stadium. VoiceFonts open up a whole new world of fun over the Internet.

Netscape has also made a strategic investment in Voxware. In return, Voxware provides some of the principal technology components for LiveMedia that will allow telephone products from multiple vendors to talk to one another. TeleVox users will be offered a free upgrade to the new technology.

First Impressions

Looking at TeleVox for the first time, you will notice that the designers didn't go for the "cellular phone" look. The product offers a clear functional approach to accessing its various features. This product is extremely easy to use. Within three minutes of installing the software, I had figured out how to make a call. There are six icons on the screen as well as direct access to the microphone volume and speaker adjustment.

One of the TeleVox features you might find helpful over time is the ability to engage in a text chat as well as send files between parties. My first introduction to text chats was on the "CB simulator" section of CompuServe. Several years later I was hooked on using IRC servers to engage in text chats with people all over the world. But when text chatting became a feature available with Internet telephony software products, I wasn't certain of the benefits. After all, if you could hear the other person's voice, why would you need to see text on your screen? With TeleVox I found that many people like to try out Internet telephony products, ever though they don't have a sound card and/or a microphone. About 20 percent of the conversations I have on TeleVox are done by typing rather than talking.

I frequently use Internet telephony products to "meet" friends on the Net at a prearranged time. But on occasion, people have needed to send me a word processing document or a sound file. With TeleVox, they can send the document or sound files while we are talking.

How TeleVox Works

Like some of the other products covered in this book, TeleVox uses the server-based model for directory services. So, when you start TeleVox, you are already logged into one or more of the TeleVox servers. Once you are online, you can "see" who else is around and available for conversations. One of the useful TeleVox features is call blocking. If you don't want to be disturbed while using TeleVox, you can enable the blocking feature. If someone else tries to call you, they will get a message that says you are blocking your calls. (Remember to turn off the blocking feature or you will not be speaking with many people!)

Installing TeleVox

Installing TeleVox is a pretty straightforward process. Figure 9.2 shows the TeleVox installation screen. To install TeleVox in Windows 3.1, you need to run TX16*nnn*.EXE. To install TeleVox in Windows 95 you need to run TX32*nnn*.EXE, replacing *nnn* with the correct version number.

When the set-up program starts, you need to click on Accept to accept the terms of the license agreement. The next screen asks you to select the destination directory for the program files. Either accept the default choice or enter a different directory. The set-up program will now install the files and prompt you to press an OK button when the installation is completed. The program also created a new Voxware program group.

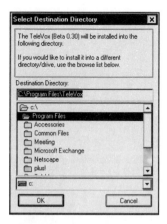

FIGURE 9.2 The TeleVox installation screen.

Product Features

When you want to call someone online, you need to see a list of current users. By clicking on the Users button from the TeleVox main screen, you'll obtain the screen shown in Figure 9.3. This is a screen shot of the TeleVox phone book. When using TeleVox, it's a good idea to keep this screen displayed. The listing of names does not update automatically. To get an updated list of names, you need to click on the Refresh button. You can request an updated list every five seconds.

Figure 9.4 shows the TeleVox Effects screen. You can display this screen by clicking on the Effects button on the main TeleVox screen. When you're in the

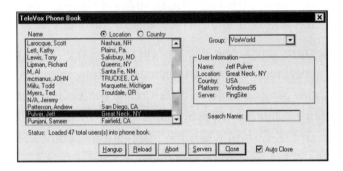

FIGURE 9.3 The TeleVox phone book.

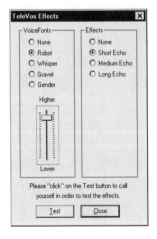

FIGURE 9.4 The TeleVox Effects screen.

middle of a conversation and have a good connection, try using the TeleVox VoiceFonts, one of the choices in Effects. Not only can you add an echo to your voice, but you can make yourself sound like a robot, change the gender of your voice, and even make it seem like you're whispering. On the other hand, if someone is having trouble understanding you, I'd recommend that you don't use any VoiceFonts.

When you click on the Send button on the main TeleVox screen, a screen like the one in Figure 9.5 will be displayed. You can use the TeleVox File Transfer Send feature to send the person you're speaking with a file from your system.

The Text button is shown in Figure 9.6. Click on this button when you want to engage in a text-chat session while using TeleVox.

Figure 9.7 shows the options you can set on TeleVox using the Options Configuration screen. From the Help bar on the main TeleVox screen, select

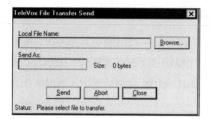

FIGURE 9.5 The Send dialog box.

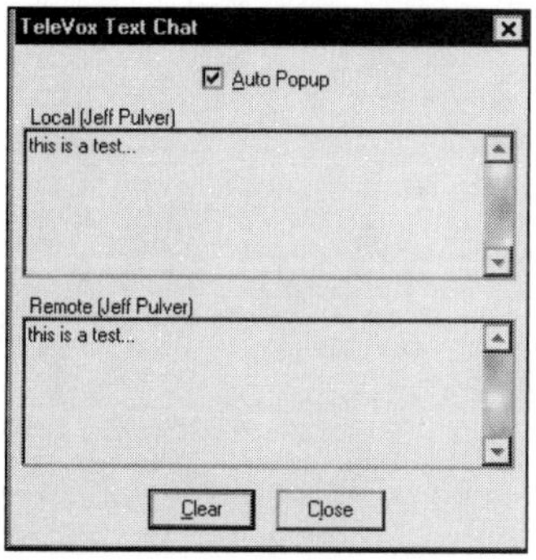

FIGURE 9.6 The Text dialog box.

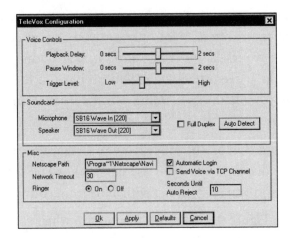

FIGURE 9.7 The TeleVox Options Configuration screen.

Network, and you'll see the information displayed in Figure 9.8. If you are interested in TeleVox's network performance, you can view the network statistics by clicking on Options Statistics, as shown in Figure 9.9.

One of the nice things about TeleVox is that you can directly control both the microphone levels and speakers via the main screen. Other products sometimes hide these two essential features, but with TeleVox it's right out in the open and easy to access.

Some TeleVox features include the following:

- An innovative interface
- Full-duplex transmission
- File-transfer capabilities

```
┌─ TeleVox Network Information ──────────────── ✕ ┐
│                                                  │
│   Highest Version:    1.1     IP Address: 204.7.54.180
│   Expected Version:   1.1     Port:       12370   │
│   Max Sockets:        256                         │
│                                                   │
│        Microsoft Windows Sockets Version 1.1.     │
│               Running on Windows 95.              │
│                                                   │
│                  ┌───────────┐                    │
│                  │    Ok     │                    │
│                  └───────────┘                    │
└───────────────────────────────────────────────┘
```

FIGURE 9.8 The TeleVox Network Information screen.

FIGURE 9.9 The TeleVox Options Statistics screen.

- Text chat
- Caller ID
- Call blocking
- Automated login/logout
- User-defined group
- Netscape interface to web pages associated with "announcements"

At the bottom quarter of the TeleVox screen, you may notice a small "announcement." If you are using Netscape 2.x or greater, you are linked to a "web space" that links you to a particular URL or website. If Voxware decides to deploy the "advertising" model as WebPhone has done, the company already has a bulletin board in place with this feature.

Where to Find TeleVox

A free version of TeleVox is always available from the Voxware's website. The URL is http://www.voxware.com.

Usage Tips

As the VON marketplace matures, the "minimum" system requirements keep moving upward. TeleVox is no exception to this general rule. The product will run on a 486/66 with an 8-bit sound card and a 9.6K connection to the Net. However, for the greatest success, it is recommended that you run TeleVox on a Pentium-class machine with a 16-bit sound card and a 28.8K Net connection.

Some of the configuration options for TeleVox include the following:

- Playback delay
- Pause window
- Trigger level
- Groups

Playback Delay

Playback delay is an option that controls the amount of time before speech is played back. If your conversations sound choppy, you might want to increase the playback delay value. The default value is one second, and can range from 0 to 2 seconds. To change the default value, click on Options, and under "Voice Controls" adjust the slider switch for playback delay.

Pause Window

Pause window is a setting that controls how quickly TeleVox stops transmitting when you finish speaking.

Trigger Level

Trigger level is a setting that allows TeleVox the ability to adjust to the sensitivity of different microphones. This is a rather useful feature when you are trying out various microphones.

Groups

Like some of the other server-based telephony products, TeleVox allows you to create private chat areas for yourself and other TeleVox users on a server.

No other TeleVox users will know that the group exists unless you tell them about it. If you know the group name, you can add it to your group list and then access the other users in that chat area. An improvement to this feature would be to include password-protect access to these private chat areas.

TeleVox directory services are set up so that everyone who logs on is placed in the same group, called "VoxWorld." Voxware has a fundamental understanding of its user community. The creation of the "family" group is the vendor's way of suggesting to everyone who uses TeleVox that if they decide to join the family group, they need to respect the "family values" represented in that group. Other Internet telephony products such as FreeTel attract a user community that uses Internet telephony for adult-related (often X-rated) services.

Chapter 10

FreeTel

FreeTel, an Internet telephony product from FreeTel Communications (see Figure 10.1), was the first Internet telephony product to feature sponsorship (in other words, advertising) as a primary focus of its product.

Once you start using FreeTel, you might start to get the feeling "while you talk—here's a word from our sponsor." This is a very novel approach to marketing, at least on the Net. When you click on an ad, FreeTel will launch your browser and bring you to a home page associated with the advertising. If you don't want to see the ads, all you need to do is pay for the product; how to do so is described at the end of this chapter.

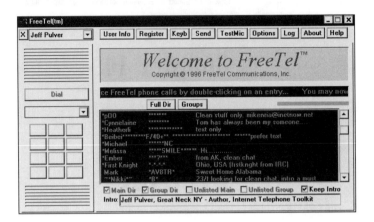

FIGURE 10.1 The FreeTel Main screen.

Like many other Internet telephony products, FreeTel uses a server-based model. At first glance, you can tell that FreeTel was designed to look like a telephone. And like other products, FreeTel also supports both file transfer and text-chat mode. The file transfer feature is useful, but I've yet to find a real need for the text-chat mode, as I mentioned in Chapter 9. One of the features you may find helpful over time is the ability to engage in a text chat as well as send files to the person you're speaking with.

FreeTel helps to push upward the minimum system requirements for Internet telephony software. When I tried using FreeTel on my 486/50 computer and a 14.4K modem, I experienced a lot of "choppy" conversations. Others I spoke with also said that FreeTel used a lot of their systems' resources and that it performed much better on a faster computer. In fact, when I tried FreeTel on my friend's Pentium (P100) computer, the choppiness was gone, and each conversation sounded very good. With this in mind, you might want to consider running FreeTel on a Pentium-class machine with a 16-bit sound card and a 28.8K connection to the Net. Although the product will run on a 486/66 computer with a 16-bit sound card and a 14.4K connection, the quality is drastically inferior.

First Impressions

Looking at FreeTel, you will notice that the designers decided to add a telephone keypad to the front of the product. After using Internet Phone and WebPhone, I felt that FreeTel had visible features that were influenced by both of these products. This product also offers a clear, easy-to-use, functional approach to accessing its various features The buttons on the screen trigger the following functions:

- User information
- Directory
- Test microphone
- Options
- Log
- About
- Help
- Dial

In addition to these buttons, there are twelve blank buttons representing a telephone keypad. The microphone volume and speaker adjustment are located behind two sliding panels. FreeTel is currently the only Internet telephony product that provides a visual indication of whether or not the person you want to call is engaged in a conversation before you attempt to contact them. All of the other Internet telephony products provide indicators such as a busy signal or a dialog box telling you the party is busy, but after you dial a number. On FreeTel, asterisks appear to the left of users' names if they are already speaking with someone.

How FreeTel Works

Like some of the other products covered in this book, FreeTel uses the server-based model for directory services. So, when you start FreeTel, you are logged into the company's servers. Once you are online, you can "see" who else is around and available for conversations by going to the main FreeTel screen and checking out the scrollable listing of the people who are online and using Free-Tel. No need to click on any icon—all you need to do is scroll to the person's name, click on the Dial button, and you will be able to communicate with them.

Installing FreeTel

To install FreeTel, put a copy of FT100.EXE in a temporary directory such as C:\TEMP. From C:\TEMP, run FT100.EXE. FT100.EXE will unpack nine files. Once the files have been unpacked, run SETUP.EXE, which is now in the directory in which you placed FT100.EXE.

The set-up program will ask you for the name of a directory into which you want to install FreeTel. I recommend selecting the default, C:\FREETEL. If you are using full-duplex SoundBlaster drivers, you should click on the box that is shown in Figure 10.2.

The set-up program will also update your SYSTEM.INI file after saving a back-up copy as SYSTEM.SAV. Then the program creates a Windows program group called FreeTel. The installation program will automatically restart Windows.

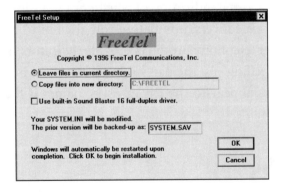

FIGURE 10.2 FreeTel Installation screen.

To start FreeTel from Windows, simply click on the FreeTel icon in the Free-Tel program group.

Product Features

FreeTel's features include the following:

- Full-duplex operation. FreeTel comes with its own set of full-duplex drivers for owners of the Creative Labs SoundBlaster 16 card. In addition, FreeTel fully supports full-duplex operation with all other full-duplex sound cards.

- Electronic phone directory. When you are online, FreeTel provides an online display of everybody who is logged onto the FreeTel server.

- Advanced Caller ID. Before you actually make a call by double-clicking on an entry or on the Dial button, you can type in an electronic introduction message—something like, "Hi, I'm Joe, would you like to talk about running?"—and your message will be displayed on the other party's screen when FreeTel is ringing. Based on the introduction message, the person you are calling can decide whether or not to accept your call.

- Phone web links. With FreeTel you can allow people to start a Free-Tel call to you from your home page. You can also keep a private web page of people you speak with on FreeTel. If you want to set up a link on your home page to call your friend Steven Bielik, you would place the link in this form:

```
<a href="FreeTel: Steven Bielik>Call Steven Bielik<a/>
```

- Once your link is set up, a user looking at your home page can click on a button marked "Call Steven Bielik," and your copy of FreeTel will automatically place a call to Steven. If Steven is not online, the caller will be notified that he is unavailable.

- Keyboard communicator. You can use FreeTel to communicate with someone via your keyboard in addition to talking with them.

- Booster feature. When there is network congestion, you can enable FreeTel's booster feature, which—although increasing the delay—improves the sound quality. The booster feature switches the communication between two parties from the UDP protocol to a TCP protocol. The Booster button is located near FreeTel's microphone controls but is visible only when you're in a conversation with someone.

- File transfer capability. With FreeTel you can send files to the person you are speaking with.

You can manually override the voice activation feature (VOX) by pressing and holding the CTRL key. However, this feature works only if the other person is using voice activation, too.

Another feature is that the product is available at no cost and will continue to be free. However, if you grow tired of seeing the FreeTel advertising while you're online, for a fee of $29.95 you can order FreeTel+. FreeTel+ has all of the features of FreeTel, plus private group capability and an unlisted name capability. With FreeTel's server-based model, all of the topics (i.e., chat rooms) into which an unregistered user can enter are public. This means that just as you can see the names of everybody who is in each of the rooms, so can everybody else. For people who purchase the registered copy of Free-Tel, one of the benefits is that you can create private rooms. The value of the unlisted name capability is that nobody can call you unless they know your name.

Where to Find FreeTel

A free version of FreeTel is always available from the FreeTel Communications website. The URL is http://www.freetel.com.

Chapter 11

CoolTalk

CoolTalk from Netscape Corporation (see Figure 11.1) might seem similar to the other Internet telephony products discussed in this book, but there is a big difference: it's bundled with the most popular web browser in the world. When Netscape purchased Insoft, a small Internet multimedia company, a few things happened:

- Insoft got the instant recognition it deserved for developing and marketing leading-edge corporate communication tools such as its whiteboard products and, most recently, its Internet telephony product, CoolTalk.

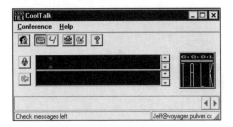

FIGURE 11.1 The CoolTalk Main screen.

- Netscape may have put fear into the hearts of the relatively smaller Internet telephony software companies, because once CoolTalk becomes part of Netscape 3.0, which will be used by several million (if not more) people practically overnight, the entire dynamic of the Internet telephony marketplace will change forever. In fact, it was just a couple of months after Netscape 3.0 was announced that Microsoft joined the Internet telephony marketplace with the introduction of its own product, NetMeeting (discussed in Chapter 12).

With the announcement of CoolTalk, Netscape was also put into the powerful position of helping to move the marketplace toward agreement on standards. Netscape held a press conference in late January 1996 and announced its support of the RTP protocol and its LiveMedia initiative. Between the efforts of Netscape and those of IBM, I believe there will be a set of commercial standards adopted and put into production by the majority of the Internet telephony software developers.

What Is CoolTalk?

CoolTalk is a real-time desktop audio conferencing and data collaboration tool specifically designed for the Net. CoolTalk provides real-time audio conferencing at 14.4K or 28.8K modem speeds and includes a full-function whiteboard, a text-based chat tool, and an answering machine.

CoolTalk runs on a 486-class PC with 8MB of RAM, running MS Windows 95, NT, or 3.1, but versions also exist for UNIX: Solaris 1, Solaris 2, HP-UX, Irix, or Digital Unix. This means that once Netscape 3.0 ships, CoolTalk will become the most popular multiplatform Internet telephony product. I'm sure that there will be a Mac version of CoolTalk in the not-too-distant future.

CoolTalk and its technology will serve as the basis for LiveTalk, an RTP-based application that uses the LiveMedia framework. TeleVox is another product that is part of the LiveMedia framework.

First Impressions

The interface for CoolTalk is easy to use, and by using the LIVE.NETSCAPE .COM directory service server, you will be able to pick any number of people to speak with. One of my earliest collaborative computing experiences was with another CoolTalk user on the West Coast. We ended up playing a game of tic-tac-toe on the whiteboard!

How CoolTalk Works

Like some of the other products covered in this book, CoolTalk uses the server-based model for directory services. CoolTalk refers to this model as its 411 servers. When you start CoolTalk, you are logged into one or more of the company's servers. Once you are online, you can "see" who else is around and available for conversations. One of the neat features is the answering machine, which allows you to leave a voice-mail message if you don't find the person you want to speak with.

Installing CoolTalk

It took me less than two minutes to install CoolTalk. The installation process is very straightforward. CoolTalk is installed automatically as part of Netscape Navigator 3.0. Follow these steps to set up CoolTalk:

1. When Netscape Navigator is installed, you are prompted with the question, "CoolTalk extends the Navigator with real-time telephone, data, and chat capabilities. Would you like to install CoolTalk?" Select Yes.

2. You will be asked, "Would you like to enable the CoolTalk Watch-Dog?". If you are a corporate user and have a permanent connection to the Internet, I would suggest you choose Yes. If you access the Internet with a dial-up connection, I would suggest you choose No.

After Netscape completes the installation process, it creates a Navigator program group. Double-click on CoolTalk to launch it. Now, start the CoolTalk setup wizard, which takes you through the following screens:

- Selection of modem speed
- Detection of audio devices installed on your computer
- Playback of CoolTalk audio messages
- Test of your microphone and setting
- System performance test
- Enter specific information for the CoolTalk business card

When each screen appears, press the Next button to proceed.

The information on the CoolTalk business card includes the following:

- Login
- Name
- Title
- Company
- Address
- Phone
- Fax
- E-mail

- Optional .BMP file containing your photo

When you first establish contact with another person using CoolTalk, the program automatically exchanges business cards with the person you just called. I encourage you to enter your actual name, title, and company, but for your own security, I would not suggest that you enter your actual phone number.

The CoolTalk WatchDog

You might want to know how much bandwidth is used by a CoolTalk session. The answer depends on the audio codec you are using and the amount of whiteboard data (if any) that you send. Audio can occupy 2.4K (RT-24), 7K (5KHz GSM), or 15K (8KHz GSM) of bandwidth.

When CoolTalk is installed with Windows, a program called CoolTalk WatchDog is also installed. The WatchDog is a small application that runs in the background. It registers you with Netscape's 411 directory server (if you desire) and answers incoming calls for you by automatically starting CoolTalk for you. This way you don't have to run CoolTalk all the time, but you are still available for calls.

Product Features

After a while, all of these Internet telephony products start to look the same. The feature that sets CoolTalk apart from the rest is that it allows you to communicate with any LiveMedia-compatible Internet phone.

After you've had a chance to review some of the competing telephony products, you will also start to notice that some products have features that are easy to access and can be used the first time without looking for the instructions. Netscape's CoolTalk is such a product.

Some of the features included with CoolTalk are the following:

- *Call dialog box.* Suppose you want to call billy@joel.com. With CoolTalk, all you would need to do is to enter the following:

    ```
    billy@joel.com
    ```

    in the CoolTalk Call dialog box (see Figure 11.2). CoolTalk will attempt to make a connection as soon as you press Enter.

- *Full-duplex support.* If you have a full-duplex sound card, CoolTalk supports full-duplex mode.

- *Whiteboard.* CoolTalk allows you to share a common whiteboard among a group of two or more people (see Figure 11.3).

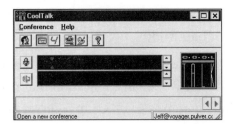

FIGURE 11.2 The CoolTalk Call dialog box.

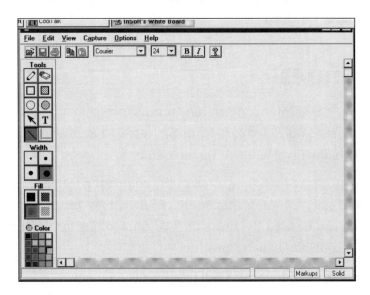

FIGURE 11.3 The CoolTalk whiteboard.

- *Text chat.* In addition to speaking with someone, CoolTalk allows you to type messages to another party as well (see Figure 11.4).

- *Automated login/logout.* CoolTalk can be configured to automatically start or stop when you turn your computer on or off.

- *Voice mail.* You can send and receive voice-mail messages with CoolTalk (see Figure 11.5). This was one of the features of CoolTalk that really got my attention. Instead of delivering voice mail to your e-mail address, CoolTalk stores it on the computer

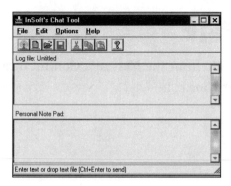

FIGURE 11.4 The CoolTalk Text Chat dialog box.

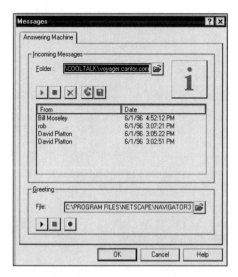

FIGURE 11.5 The CoolTalk voice-mail feature.

that's running CoolTalk. You can send a picture of yourself along with your voice-mail messages. When you initially fill out the user information, supply a patch to a .BMP picture of yourself, or any other graphic image.

- *Netscape interface to web pages associated with CoolTalk usage.* Netscape has a web page that keeps track of all of the current CoolTalk users. When you're using CoolTalk, go to the Live CoolTalk directory at: http://live.netscape.com.

Where to Find CoolTalk

A free version of CoolTalk is available from the Netscape Communications website. The URL is http://www.netscape.com.

Chapter 12

NetMeeting

When Microsoft Corporation introduced the NetMeeting product in June 1996, it established a new standard, even for Microsoft. The Beta I version of the software product, which I downloaded and tested, was among the best beta versions of any Internet telephony software product I've tested to date.

The most amazing thing about the release of NetMeeting is that it differed from most product announcements. Instead of the usual practice of announcing a product and shipping it sometime in the future, the day Microsoft announced NetMeeting was the same day it was available for download over the Internet.

If you compare NetMeeting to all of the other Internet telephony products already discussed in this book, the interface is pretty intuitive and rather easy to use (see Figure 12.1). Furthermore, the other beta testers I talked to while I was testing the software were all pretty much amazed at how the product performed. Remember, we were using the Beta I version of the software. From my past experience with packet dropoffs and hard-to-understand audio, I was impressed with the high quality of the audio I received using NetMeeting.

Using NetMeeting

The first steps in using NetMeeting are pretty simple:

1. To initiate a call with NetMeeting, make sure that you are using ULS.MICROSOFT.COM when NetMeeting prompts you for the conferencing server name.

2. After you are connected, click on the Directory button and select somebody that you would like to contact (see Figure 12.2).

Application Sharing

With NetMeeting, you can transfer files between NetMeeting users as well as participate in whiteboarding sessions and/or text chats. However, the one

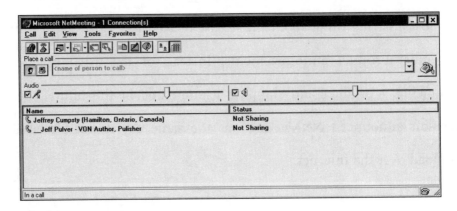

FIGURE 12.1 The NetMeeting Main screen.

FIGURE 12.2 The NetMeeting Directory button.

feature that stands out as unique is called application sharing. This means that up to 255 people can connect—either over the Internet or on a corporate intranet—and be able to view the same application. At the same time, any one of the parties can be in control of the mouse.

When you share applications, you are actually opening up *one* application on *one* computer and sharing it with everybody else who is connected, even when the connected computers do not have that application installed. This single feature will make NetMeeting a very popular product within corporate networks around the world.

Application sharing offers tremendous opportunities for collaborative computing, far beyond whiteboarding and video conferencing. The Beta I version supports only Windows 95, but Microsoft has stated a commitment to support Windows NT and Macintosh operating systems as well.

Internet Telephony Standards

Microsoft's approach to Internet telephony standards is to support two existing international standards: H.323 and T.120. This means that NetMeeting can be used immediately with the many other desktop conferencing software packages in the marketplace, including Intel's Proshare. Although it may be a long time before NetMeeting ever communicates with a LiveMedia compliant client, I believe many of the Internet telephony products discussed in this book will support H.323 and T.120 in the near future.

Whiteboard Functionality

With NetMeeting you can set up real-time audio and data conversations with groups of people. This would be a good reason to use a whiteboard (see Figure 12.3), but frankly, unless you have a specialized writing tablet, your mouse is really not a good writing instrument. For competitive reasons, the Internet telephony vendors feel compelled to provide whiteboard functionality with their products, but until such time that the standard operating environment on a PC provides writing tools, the overall value of a whiteboard is limited.

The Chat Screen

NetMeeting also allows you to perform a text chat with more than one person, as shown in Figure 12.4. While I was testing NetMeeting, I found myself using the Chat screen with people who downloaded the NetMeeting software but didn't have a microphone. Other than to help people diagnose problems with their systems, I haven't found that much use for the Chat screen feature in any of the Internet telephony software I've used.

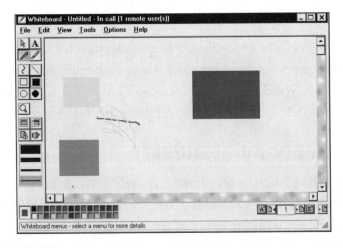

FIGURE 12.3 The NetMeeting whiteboard.

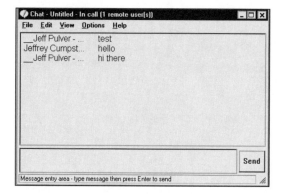

FIGURE 12.4 The NetMeeting Chat screen.

Where to Find NetMeeting

One of the interesting aspects of the software product is how the entry of NetMeeting, with its powerful conferencing features, might influence the Internet telephony marketplace. NetMeeting is currently available for free over the Internet, the audio quality is pretty good, the product really works—and it's from Microsoft. These are characteristics that point to success in the market.

NetMeeting is available for downloading at Microsoft's web page. The URL is http://www.microsoft.com/ie/conf/default.htm.

Chapter 13

Additional Voice/Video on the Net (VON) Products

In the world of voice/video on the Net (VON), the Internet really hasn't been the same since the introduction of Internet telephony. In addition to the products discussed in this book, some other Internet telephony products were recently introduced. Some of the better products include the following:

- PowWow

- VDOPhone

- SpeakFreely

- PGPhone

PowWow

Of all the utility products introduced in 1995, my personal favorite is PowWow from Tribal Voice. PowWow is a Windows product that allows up to seven people to cruise the World Wide Web together, send and receive files, and text chat with each other. The product even supports voice chatting with only specific members of the group (see Figure 13.1).

During the initial start-up period of the Free World Dial-Up project, I used PowWow to keep in touch with members of the project team. For those of you who have used IRC servers, you can think of PowWow as similar to having your own private IRC chat session, but without the need to connect with the IRC server network.

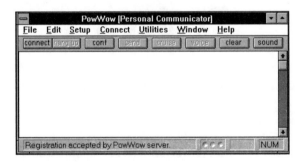

FIGURE 13.1a PowWow Main screen.

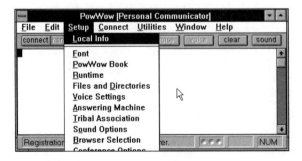

FIGURE 13.1b PowWow Main screen.

PowWow also provides support that allows you to contact someone over the Net from their web page. For example, if your e-mail address is james@tavly.com and you want to encourage people to contact you via PowWow, all you need to do is to put a link button marked as follows in the HTML of your home page:

```
powwow:jamesatalvy.com
```

VDOPhone

In March 1996, I attended the Computer Telephony Expo in Los Angeles. I had heard that VDO was going to demonstrate an alpha version of VDOPhone (see Figure 13.2), so I expected the usual product-announce- ment hoopla at the company's booth. What I wasn't prepared for was that when I got to the booth, it was empty—no gangs of marketing and sales people, no hired models. The VDOPhone demonstration was indeed live, but it consisted of a person back in VDOPhone's office who was "connected" to a PC in the booth using the VDOPhone product.

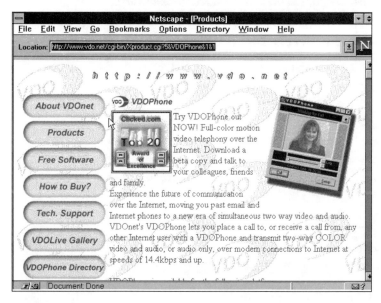

FIGURE 13.2 The VDOPhone Main screen.

At first I thought it was a setup. I saw someone on the monitor waiting to speak with someone, so I went over and started to talk into the microphone. I found out that the person slated for booth duty had to leave the show to take care of a personal emergency. Although it may not have been done on purpose, the result was an effective demonstration. At around 19.2K, there were about five to eight frames per second of video, and the audio sounded pretty good.

SpeakFreely

My friend Brandon Lucus from Tokyo has found SpeakFreely (see Figure 13.3) to have the best sounding audio of all the Internet telephony products he uses. Many others would agree. Like TeleVox and FreeTel, this product is available for free, and unlike most of the other Internet telephony products profiled in this book, both Windows and UNIX versions are available.

SpeakFreely uses the GSM codec that, according to the company's home page, reduces the bandwidth requirements to 1,700 characters per second. SpeakFreely

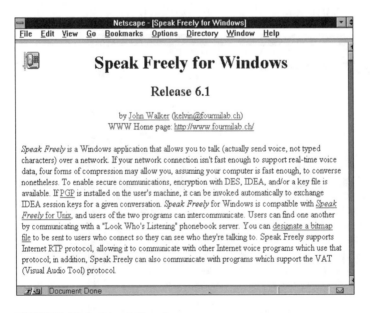

FIGURE 13.3 SpeakFreely screen.

can also communicate with other Internet telephony products that support the RTP or VAT protocols. SpeakFreely also offers voice mail and encryption.

PGPhone

PGPhone (which stands for Pretty Good Privacy Phone), shown in Figure 13.4, is one of the most secure Internet telephony products available on the Net. If you are worried about your privacy, I suggest you try PGPhone. At last check, PGPhone was available for both Windows 95 and the Macintosh, and it's possible for a Mac PGPhone user to communicate with a Windows 95 PGPhone user.

The VON Home Page

One of the pages I maintain on the Internet is the VON home page shown in Figure 13.5. The purpose of the VON home page is to provide a resource on which people can rely to stay on top of the various VON technologies and news. As new products are introduced to the VON community, they are added to the VON page.

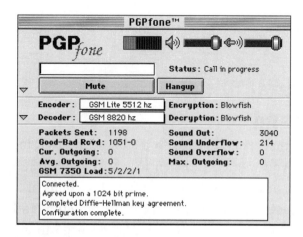

FIGURE 13.4 PGPhone's Main screen.

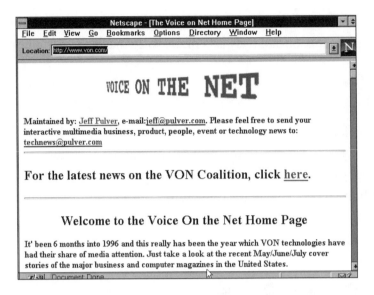

FIGURE 13.5 The VON home page.

Other VON-Related Technologies

In addition to the Internet Telephony side of Voice on the Net, other related technologies fall into the VON category. These technologies include the following:

- Audio on demand
- Video on demand
- Video conferencing

Audio on Demand

It wasn't that long ago that you could go to a site on the Net, locate the music file you thought you wanted to listen to, and spend upward of 35 minutes downloading it, only to find out that after 18 seconds of listening to it you really didn't like it. These days, with the introduction and acceptance of "streaming" technologies, all you need to do is go to a website, click on an icon, and within a couple of seconds you can begin to hear the music. If you don't like what you are listening to, you can immediately turn off the music clip and move on to something else. Today, there are even record stores on the Net that allow you to sample tracks from an album prior to purchase. Companies have been putting sample tracks of featured musical guests on the Net in order to showcase a new tune. Although it might still make news to have a new music video premiere on MTV, many promotional people from the recording industry have been going to the Net to set up websites and make sure certain sound clips are available for those who want to listen to them. These record companies recognize the great power of the Net to achieve one-to-one marketing in a one-to-many environment.

Quite a few people have taken on the role of "Internet producer," providing original content on the Net and hoping to become discovered. One such band I found is I Forget, whose home page is shown in Figure 13.6. Drop by the I Forget home page (http://iforget.com) and take a listen for yourself.

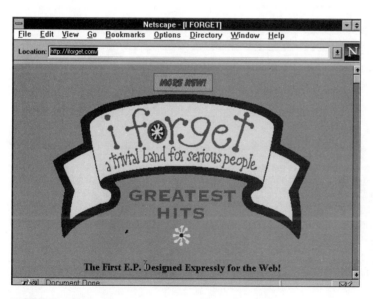

FIGURE 13.6 The Home Page for I Forget.

In addition to music sites, other websites allow you to hear a welcome message from a company's CEO, as well as sites like Bootcamp (http://www.bootcamp.com), as shown in Figure 13.7, where you can get your weekly dose of news pertaining to computers and technology, or my personal website (see Figure 13.8) where people offer personal greetings. For my home page, go to http://www.pulver.com/jeff.html.

Several companies offer products that provide streaming audio technologies. Some of the more popular products in use on the Internet today include RealAudio (see Figure 13.9), Streamworks, ToolVox, TrueSpeech, and Internet Wave.

Among the general business community, RealAudio from Progressive Networks (http://www.realaudio.com), shown in Figure 13.10, is one of the most popular audio formats you will find on the Net today. When RealAudio hit the Net in April 1995 it caused such a sudden impact that there was immediate recognition in *Time* Magazine. Streamworks from

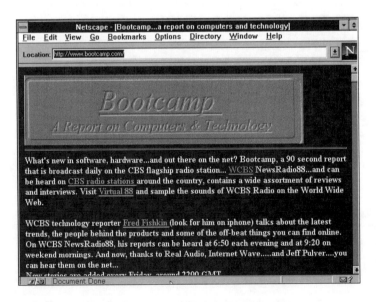

FIGURE 13.7 The home page for Bootcamp.

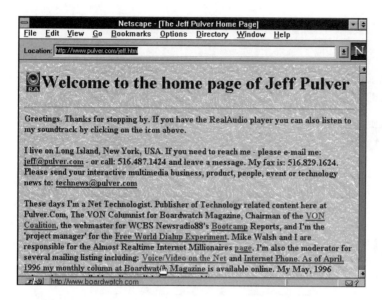

FIGURE 13.8 The Jeff Pulver home page.

Xing Technology (http://www.xingtech.com) is also very popular, especially amongst websites in the broadcast industry.

Both RealAudio and Streamworks are based on a server technology in which the original content is first encoded before it is "served" out, as shown in Figure 13.11. People who want to listen to the content of an audio file are required to download a "player," as shown in Figure 13.12. Both RealAudio and Streamworks can also be used to broadcast live events.

Recently I remember watching Bill Gates on the David Letterman show. During the classic episode, Bill was explaining to Dave about the concept of listening to a live broadcast of a baseball game over the Internet. Dave's response was something to the effect of, "Haven't you ever heard of Radio?" When Bill tried to show

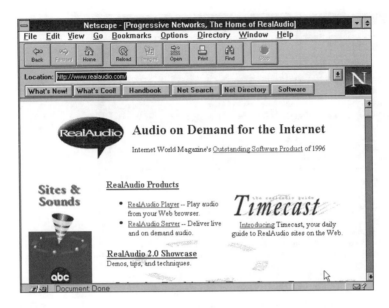

FIGURE 13.9 The RealAudio home page.

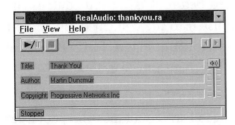

FIGURE 13.10 The RealAudio player.

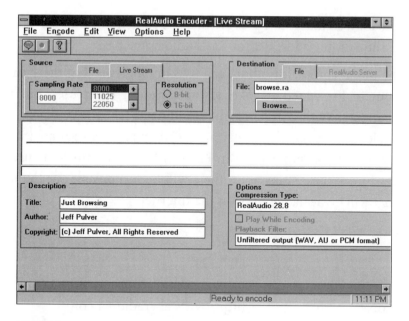

FIGURE 13.11 The RealAudio encoder.

that you can play audio on demand over the Internet, Dave's response was something like, "Haven't you ever heard of a tape recorder?" In some respects, some of us like to look at this technology the way both Bill and Dave did while at the same time trying to find some very useful applications. The subtle point here is that with this incredible enabling technology, the challenge is now to find

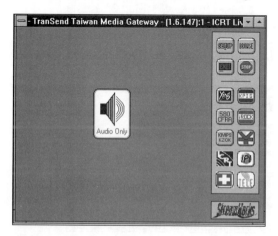

FIGURE 13.12 The Streamworks player.

unique uses for it so that it can provide valuable content for the right audience. Table 13.1 shows some addresses for RealAudio home pages.

One of the initial complaints about these technologies is that there isn't any unique content for which to take advantage of their features, but this is quickly changing. With companies like AudioNet (shown in Figure 13.13) and RadioNet and personalities like Marc Cuban and Ken Rutkowsi (home page http://www.ttalk.com, shown in Figure 13.14), you can be assured that in the near future these folks and others will be responsible for the unique content that will be available only on the Internet.

TABLE 13.1 Select RealAudio Sites

Name	Home Page
1-800 MUSIC NOW	http://www.1800musicnow.mci.com/
ABC RadioNet	http://www.abcradionet.com/
AudioNet	http://www.audionet.com/
CBC Radio	http://www.cbcstereo.com/RealTime/ soundz/realaudio/ra_menu.html

FIGURE 13.13 The AudioNet home page.

TABLE 13.1 Continued

Name	Home Page
CDNow: The Internet Jukebox	http://cdnow.com/jukebox/from=realaudio
clnet radio	http://www.cnet.com/Content/Radio/

Many major monthly PC magazines as well the weekly PC trade publications have started to put RealAudio content out on the Net. From *Hotwired* to *PC Week* to *Computerworld*, these publications now have a real "voice." The question is, how many people are listening?

There was a time when pirate radio stations were the rage across the AM band. These days, rather than taking any risks and breaking any of the FCC laws, the would-be pirate radio operators can legally set up their broadcasting facilities on the Net and even take live call-ins via the Internet telephony products that are now available. Table 13.2 shows selected Streamworks sites, where you can get information about some of these "renegades."

FIGURE 13.14 The TechTalk home page.

TABLE 13.2 Selected Streamworks Sites

Name	Home Page
KPIG Radio Online	http://www.kpig.com
Bloomberg Information News Radio	http://www.bloomberg.com/cgibin/ tdisp.sh?wbbr/index.html%20
International Community Radio Taiwan	http://www.icrt.com.tw
Internet Business Radio	http://www.ibrlive.com
Sportsline USA	http://www.sportsline.com/u/ ap/audio/index.html
Hard Radio	http://www.hardradio.com

ToolVox from Voxware (http://www.voxware.com), TrueSpeech from DSG Group Inc. (http://dsgp.com), and Internet Wave from VocalTec (http://www .vocaltec.com) all offer the ability to add streaming audio to your website at no cost. You can just download the software from these sites.

Voxware's leading-edge ToolVox codec can be found in many popular VON products, from TeleVox to CoolTalk to PowWow. In particular, the Speech-Fonts are really fun to play with.

In order to serve out ToolVox files, just as with RealAudio and Stream-works, you will need to use an encoder, as shown in Figure 13.15, to encode your files into the ToolVox format. Once encoded, all you need to do is to modify the MIME.TYPE configuration file of your web server in order to have your web server be able to server out ToolVox files on demand. The person listening will need to use the ToolVox player, as shown in Figure 13.16.

Video on Demand

In the video-on-demand field, VDOLive (http://www.vdo.net) has quickly become the standard people have gone to when they want to put "on-demand" video clips over the Internet. When I first saw VDOLive in November 1995 at

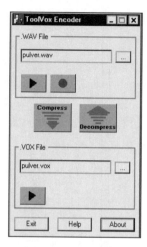

FIGURE 13.15 The ToolVox encoder.

the Fall Internet World Show, I realized I was looking at a product that would have a positive impact on the way video is delivered over the Net.

For an up-to-date listing of sites that use VDOLive, visit the VDO home page, shown in Figure 13.17.

Videoconferencing

Enhanced CU-SeeMe is a desktop video conferencing software product designed for real-time person-to-person or group conferencing. This means you can use it for one-to-one, one-to-many, and many-to-many conferences. CU-SeeMe runs over the Net or on your corporate intranet and is available for both Mac and Windows users.

Josh Quittner's technology column in my local Long Island newspaper, *Newsday*, started me on the road to tracking emerging technologies on the Internet.

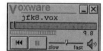

FIGURE 13.16 The ToolVox player.

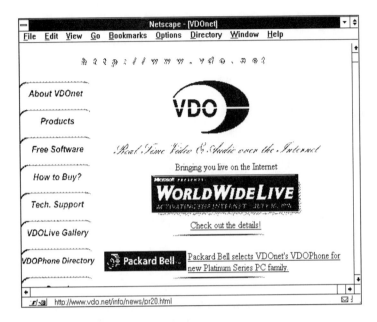

FIGURE 13.17 The VDO home page.

From the first article I read, in which Josh wrote about his experiences using CU-SeeMe, I was hooked. By the way, I strongly suggest you take a look at one of Josh's latest creations: The Netly News (http://pathfinder.com/Netly/).

It took a little time, but after spending eight months trying to use CU-SeeMe using a 14.4K dial-up connection to the Internet, I finally made a gigantic leap of faith and ordered a dedicated 56K line to my home. CU-SeeMe played a lot better on the 56K line.

When White Pine (http://www.wpine.com) took over the active development of CU-SeeMe I was immediately impressed. White Pine's Enhanced CU-SeeMe, shown in Figure 13.18, has set the standard for two-way video conferencing on the Internet. If you try only one desktop video conferencing product, you should download Enhanced CU-SeeMe or play with the version that is included on the CD-ROM that accompanies this book.

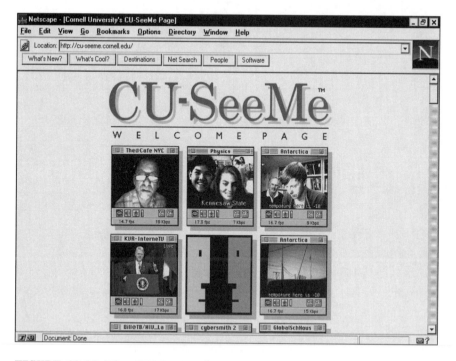

FIGURE 13.18 The CU-SeeMe home page.

Enhanced CU-SeeMe as well as the other VON technologies will continue to improve and provide additional advanced functionality. Over time, these and other products will start to redefine how we use the Internet as an international communications tool.

Appendix A

Internet Telephony Resource List

The list included in this appendix is from my personal collection of information on vendors of Internet telephony products, add-on sound/video products, and sound cards. If you are aware of additional companies that should be listed, e-mail me at jw-questions@pulver.com with the subject, "Add to List." I would also be interested in hearing your experiences with these products.

Internet Telephony Vendor Directory

3D Voice Chat

OnLive! Technologies

10131 Bubb Road

Cupertino, California 95014

Phone: (408) 366-6000

Fax: (408) 366-0357

http://www.onlive.com

CoolTalk

Netscape Communications Corp.

501 E. Middlefield Rd.

Mountain View, CA 94043

Phone: (415) 937-2555

Fax: (415) 528-4124

http://live.netscape.com

DigiPhone

Third Planet Publishing, Inc.

P.O. Box 797728

Dallas, TX 75379

http://www.planeteers.com

FreeTel

FreeTel Communications, Inc.

540 N. Santa Cruz Ave, Suite 290

Los Gatos, CA 95030

Fax: (408) 358-6385

http://freetel.inter.net/freetel

Global Chat

Quarterdeck Corp.

13160 Mindanao Way

Marina Del Rey, CA 90292-9705

Phone: (310) 309-3700

Fax: (310) 309-4217

http://www.qdeck.com/chat/

IBM Internet Connection Phone
IBM Corp.
Boca Raton, FL
http://www.ibm.com/internet/icphone.html

InterCom
http://revsoft2.is.net/ic/intercom.html

Internet Phone
VocalTec Corp.
35 Industrial Parkway
Northvale, NJ 07647
Phone: (201) 768-9400
Fax: (201) 768-8893
http://www.vocaltec.com

IRIS Phone
http://alpha.acad.bg/iris

NetMeeting
Microsoft Corp.
One Microsoft Way
Redmond, WA 98052-8303
http://www.microsoft.com/ie/conf/default.htm

Net2Phone
IDT
294 State Street
Hackensack, NJ 07601
Phone: (201) 928-1000
Fax: (201) 928-1057
http://www.net2phone.com

PGPhone

Phil Katz

MIT

77 Massachusetts Avenue

Cambridge, MA 02139-4307

Phone: (617) 253-1000

http://web.mit.edu/network/pgpfone

PowWow

Tribal Voice

627 West Midland Avenue

Woodland Park, Colorado 80863-1100

Fax (719) 687-0716

http://www.tribal.com

Speak Freely

John Walker

http://www.fourmilab.ch/index.html#speakfree

TeleVox

Voxware, Inc.

305 College Road East

Princeton, NJ 08540

Phone: (609) 514-4100

Fax: (609) 514-4101

http://www.voxware.com

TS Intercom

Telescape Corp.

Vancouver, British Columbia Canada

http://www.telescape.com

VDOPhone

VDOnet Corp.

4009 Miranda Ave. Ste. 250

Palo Alto, CA, 94304
Phone: (415) 846-7700
Fax: (415) 846-7900
http://www.vdolive.com/vdophone

WebPhone
Netspeak Corp.
902 Clint Moore Road
Suite #104
Boca Raton, FL 33487
Phone: (407) 997-4001
Fax: (407) 997-2401
http://www.itelco.com

WebTalk
Quarterdeck Corp.
13160 Mindanao Way
Marina Del Rey, CA 90292-9705
Phone: (310) 309-3700
Fax: (310) 309-4217
http://talk.qdeck.com

Add-On Sound/Video Products

Once you have set up your Internet telephony environment, you might want to consider one or more of these accessory products.

Transphone
The Firecrest Group plc
5 Stratford Place
London, England W1N 9AE
Phone: (171) 409-1214
Fax: (171) 409-1209
http://www.transphone.co.uk

SoundXchange

InterActive, Inc.

204 North Main

Humboldt, SD 57035

Phone: (605) 363-5117

Fax: (605) 363-5102

http://www.iact.com/sxc.html

Jabra

Jabra Corp.

9191 Towne Centre Drive

Suite 330

San Diego, CA 92122

Phone: (619) 622-0764

Fax: (619) 622-0353

http://www.jabra.com

QuickCam

Connectix Corp.

2655 Campus Drive

San Mateo, CA 94403

Phone: (415) 571-5100

Fax: (415) 571-5195

http://www.connectix.com/connect/CPM.html

NexPhone

InterNEX Technologies, Inc.

2994 Scott Blvd

Santa Clara, CA 95054

Phone: (408) 727-6584

Fax: (408) 727-5508

http://www.inext.com

Sound Cards

In addition to contacting these vendors, you will find this sound-card website helpful: http://www.wi.leidenuniv.nl/audio.

Acer
2641 Orchard Parkway
San Jose, CA 95134
Phone: (408) 432-6200
Fax: (408) 922-2933
http://www.acer.com

Creative Labs
1523 Cimarron Plaza,
Stillwater, OK 74075
Phone: (405) 742-6622
Fax: (405) 742-6633
http://www.creaf.com

Gravis
101-3750 North Fraser Way
Burnaby, BC Canada V5J 5E9
Phone: (604) 431-5020
Fax: (604) 431-5155
http://www.gravis.com

Turtle Beach Systems
5690 Stewart Ave.
Fremont, CA 94538
Phone: (510) 624-6200
Main fax: (510) 624-6291
http://www.tbeach.com

Additional Vendors and Industry Resources

The following is a listing of companies that are involved in other aspects of the voice/video on the Net (VON) industry. Under each of the categories you will find the name of the product and the URL to the vendor's home page. For the most up-to-date information on the companies listed below, visit their home page.

Audio-on-Demand Products

Audio-on-demand products provide "streaming" audio to websites. In other words, when you click on an icon, within a second or two, you are able to hear a sound clip. You don't have to wait for the entire segment to be downloaded. Many sites on the Internet—from radio stations to record stores to Fortune 500 Corporations—take advantage of this technology. In order to listen to these streaming audio products, you need to download what is called a "player" for each of these respective companies' websites.

RealAudio
Progressive Networks
http://www.realaudio.com

Streamworks
Xing Technology
http://www.xingtech.com

Internet Wave
VocalTec, Inc.
http://www.vocaltec.com

TrueSpeech
DSP Group, Inc.
http://www.dspg.com

Voxware

Voxware

http://www.voxware.com

Accoustic Player

http://www.ntt-ad.co.jp/ntt-powernet/vol1/audiolink/index_e.html

DirectAudio Player

http://www.cam.org/~noelbou/gsm_wine.html

NetSound

http://sound.media.mit.edu/~mkc/netsound.html

Live Audio Products

The products listed here are used on the Internet to produce live events like conferences or seminars.

RealAudio 2.0

http://www.realaudio.com

Precept—Flashware and Internet TV

http://www.precept.com

Streamworks

http://www.xingtech.com

Additional Audio Products

Audio Wave

http://www.audiowav.com

DialWeb

http://www.dialweb.com

VisualIRC

http://apollo3.com/~acable/virc.html

Voice E-Mail

http://www.bonzi.com

Internet Call
http://dsp.ee.cuhk.edu.hk/proj/icalldl.html

Live Video Products

The following companies provide players that allow you to receive live video over the Internet. You'll also find a helpful website at http://www.nrv.net/~ipr/ivng.

Enhanced CU-SeeMe
White Pine Software
http://www.wpine.com

Flashware and Internet TV
Precept
http://www.precept.com

FreeVu
FreeVu Communications
http://www.freevue.com

Streamworks
Xing Technology
http://www.xingtech.com

Vosaic
http://www.uiuc.edu/ph/www/vosaic

Video-on-Demand Products

Streamworks
http://www.xingtech.com

VivoActive—Player, Producer, and Rehearsal Server
http://www.vivo.com

VDOLive
http://www.vdo.net

Video-Conferencing Products

Cinecom

http://cinecom.com/CineCom/cinvdrct.html

Enhanced CU-SeeMe

http://www.wpine.com

FreeVue

http://www.freevue.com

VDOPhone

http://www.vdolive.com/vdophone

VidCall

http://access.digex.net/~vidcall/vidcall.html

Worldgroup

http://www.gcomm.com

Audio/Video Plug-Ins for Netscape

If you use the Netscape browser, Version 2.0 and above, you might enjoy trying the following plug-ins. In the future similar products might support ActiveX from Microsoft and Inferno from AT&T.

Action

Open2U

http://www.open2u.com/action

CoolFusion

Iterated System

http://www.iterated.com

Crescendo Plus

Live Update

http://www.liveupdate.com

Echospeech
Echo Speech Corp.
http://www.echospeech.com

Koan by SSEY
http://www.sseyo.com

MacZilla
Knowledge Engineering
http://maczilla.com

BeingThere
Intelligence at Large
http://www.beingthere.com

PreVu
InterVu
http://www.intervu.com

RapidTransit
FastMan
http://monsterbit.com/rapidtransit/

RealAudio
Progressive Networks
http://www.realaudio.com

Talker
MVP Solutions
http://www.mvpsolutions.com

ToolVox
Voxware, Inc.
http://www.voxware.com

VDOLive
VDONet
http://www.vdo.net

ViewMovie
Ivan Cavero Belande
http://www.well.com/~ivanski/

VivoActive Player
Vivo Software
http://www.vivo.com

Additional Multimedia Products

Adobe Premiere
Adobe
http://www.mv.us.adobe.com/Apps/Premiere.html

Precept Flashware and Internet TV
Precept
http://www.precept.com

Shockwave
Macromedia
http://www.macromedia.com

VocalTec
Insitu Conference
http://www.insitu.com

EMULive
http://www.jcs-canada.com

Phylon Playlink
Phylon
http://www.phylon.com

Voyager CDLink
http://www.voyagerco.com/cdlink/cdlink.html

Snappy
Play Inc.
http://www.play.com

I-Animate
http://www.tekweb.com/ianimate.html

Asymetrix Web 3D
http://www.asymetrix.com

Net Tool
http://www.duplexx.com

Sizzler
http://www.totallyhip.com

Publishers Depot
http://www.publishersdepot.com

Microsoft ActiveX Conferencing

http://www.microsoft.com/intdev/msconf/msconf.htm

Select VON Content Providers

If you decide to experiment with some of the audio- or video-on-demand products, the following sites are providers of interesting audio and video content.

ISP-TV Network
http://www.digex.net/isptv

PC Week Radio
http://www.pcweek.com/radio

@Computerworld Minute
http://www.computerworld.com

net.radio
http://www.audionet.com/pub/netradio/test.htm

Fred Fishkin Bootcamp Reports
http://www.bootcamp.com

I Forget
http://iforget.com

Live Update's Cool Site of the Day
http://www.tstmedhat.com/~neburton

ABC Radio Network
http://abcradio.ccabc.com

Bloomberg
http://www.bloomberg.com

Tech Talk Radio Network
http://www.ttalk.com

TechBabble
http://www.techbabble.com

The Devoid Corp.
http://www.ozemail.com.au/~devoid

Internet-to-Fax Products and Projects

The following sites provide information on how you can send a fax to another location over the Internet.

Internet Fax/Phone Gateway
http://fax.phone.net/faxfaq.html

JFAX—Fax and Voice-Mail Network
http://www.jfax.net

The VON Mailing List

If you would like to stay on top of the latest breaking news in the VON industry, you should take a moment and subscribe to the VON mailing list. This mailing list is dedicated to the broader discussion of all of the various VON products and technologies.

To subscribe, send e-mail to majordomo@pulver.com, leave the subject blank, and in the body write "subscribe von-digest."

Appendix B

Free World Dial-Up (FWD) FAQs

In February 1995, a small U.S. company, VocalTec, launched the Internet's first viable telephony software product, Internet Phone, which now allows thousands of computer users across the globe to make real-time telephone calls to one another over the Internet for free, using only a PC running Windows and equipped with a sound card, microphone, and speakers. Seven months later, a group of Internet Phone hobbyists launched the next stage of the telecommunications revolution. The Free World Dial-Up Global Experiment was a grassroots, completely global, noncommercial venture that extended the power of Internet telephony to ordinary people through local telephone lines.

This appendix presents some answers to the most frequently asked questions (FAQs) regarding the FWD project.

What Is FWD?

Free World Dial-Up is a grassroots global experiment that allows Internet users to make telephone calls not just to others on the Internet, but to any telephone number in local calling areas across the globe.

Using popular Internet telephony software such as Internet Phone or DigI-Phone, users contact a remote server in the destination city of their call. This server "patches" the Internet phone call to any phone number in the local exchange. The system is regulated by a client/server structure and software that our group is developing in several geographical areas of the world.

As an example of what we envisioned, a user in Hong Kong can use an Internet-based server in Paris to effectively dial any phone number and talk with anyone in the Paris area for free. No longer are free phone calls limited to the technical elite!

This service is offered to the public for free for exclusive noncommercial use, solely as a means for friends and family around the world to stay in touch. The purpose of this project is to prove to the world that it can be done.

The first stage of this experiment enabled users of the Internet and Internet telephony software to dial out to regular phones. In the future, "dial-in capability" may also be added. This would allow people to call into this system from any touch-tone telephone, creating a seamless system in which users on neither side of the connection need to have access to the Internet or a computer.

How Did this Project Get Started?

On the IPhone mailing list I moderated, Internet Phone enthusiasts in a group forum were toying with the idea of just such a global telephony service for months. It wasn't until October 1995 that Izak Jenie in Jakarta and Steven Mercurio in New Jersey sparked widespread interest in launching this experiment when they announced they were working on implementations of the Internet-to-telephone pitch.

Brandon Lucas in Tokyo crystallized interest in this project a few weeks later with a posting about monopolies and high communications prices. Within hours of that posting, the response grew so strong that I worked with Brandon in Tokyo to name and launch the new experiment: Free World Dial-Up (FWD).

What Are All the Parts of this System?

We outlined the following parts for our FWD system:

- FWD client, to run on the caller's machine
- FWD server, to run on the computers that link the Internet to local calls
- Global server, which will keep a list of all servers and the real-time status of each one

What Platform(s) Will this System Run On?

The FWD client and server are now being developed for Windows. In the future, we would like to port the software to the Macintosh platform, but we need Mac developers to work with us on this expansion. The FWD global server will be designed to run on UNIX, and in the future, we hope to have it running on Windows NT and Linux.

What Is the Timeframe of this Project?

We launched the FWD project in October 1995. Development of the client and server software has already begun, and the first beta test of this world-wide, free service was released at the end of 1995. We expect to open this service to the public in late 1996, after we build up a strong server network and iron out the bugs.

While the FWD project may end before the end of 1996, the spirit of the experiment will continue.

How Is this Experiment Organized?

I coordinate the project from my home in New York. We currently operate two mailing lists, one for general discussion and the other for software/hardware developers. You can sign up for the general list by sending e-mail to majordomo@pulver.com and leaving the subject blank. Write in the body of the message, "subscribe free-world-dialup@pulver.com."

In terms of administration, please see the next section.

Who Is Coordinating the Experiment?

The coordinators of the FWD Experiment are as follows:

- Jeff Pulver in New York City is the global project manager and is also coordinating technical issues among developers.
- Izak Jenie in Jakarta is coordinating the protocol design/program design of the FWD system.
- Lynda Meyer in New York City is the legal coordinator.
- Brandon Lucas in Tokyo is coordinating server administrator and business liaison.

There are several others working on this project who prefer to remain anonymous.

What Kind of Participation Do You Need for the Project?

Basically, we need assistance from volunteer servers.

Volunteer Servers

Anybody can volunteer their computer to become a server at little financial cost! And we need participation from all over the world. To offer your computer as a server, you will need one connection to the Internet (either dedicated or dial-up) and one separate phone line to patch calls to your local calling area. That means that if you are on a dial-up connection to the Internet, you need a total of two phone lines. You will also need a Windows machine and sound card as well as a voice modem that uses the Cirrus Logic chipset to patch through the calls. We will, of course, give you all the necessary software for free.

Where Can I Get a Voice Modem?

Please see the list of voice modems now available in Appendix A. I include a list on my website (link to http://www.pulver.com/fwd).

How Have Phone Companies Reacted to this Project?

In less than one week after the Free World Dialup Experiment was announced, ACTA, the Americas Carriers Telecommunications Association, filed a petition with the FCC to have the sale and use of Internet telephony products banned in the United States. In June 1996 at the IETF meeting in Montreal, Blair Levin, FCC Chief of Staff, delivered a message on behalf of FCC Chairman Reed Hundt, which stated ". . . the right answer at this time is not to place restrictions on software providers or subject Internet telephony to the same rules that apply to conventional circuit switched voice carriers." These words were a major victory for both the VON Coalition and anybody who wanted to take part in the Free World Dialup Experiment. Our emphasis was never on destroying telecommunications companies but rather on making it easier for average people to communicate.

Isn't It Illegal?

Because this experiment is for hobbyists in a noncommercial environment, we don't believe that we will confront legal difficulties in the United States. We do imagine that this project will raise eyebrows in several other countries, especially those that encourage and support telecommunications monopolies. It will be interesting to see how they react to this new development.

In the future as other companies provide commercial versions of this service, we expect that some regulatory agencies will try to tariff the person who owns the hardware that does the interconnection between the Public Switched Telephone Network (PSTN) and the Internet.

We are researching several legal and regulatory aspects for this project. If you would like to join in the debates, please make sure to sign up on the list.

Won't the Telecoms Stop the Internet Because of this Project?

We are researching this aspect of the project, but we suspect that any trend toward unlimited calling will create a larger market for the telecom giants, and that they will benefit handsomely from economies of scale, even at Internet prices.

Where Can I Send My Questions?

If you have any questions regarding the Free World Dial-Up Project, please send e-mail to jeff@pulver.com. In the subject, write "FWD-Question."

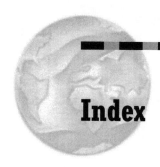

Index

CUSTOMER NOTE: